Toll-like Receptors

Edited by Nima Rezaei

Published in London, United Kingdom

IntechOpen

Supporting open minds since 2005

Toll-like Receptors
http://dx.doi.org/10.5772/intechopen.80367
Edited by Nima Rezaei

Contributors
Nibaldo C. Inestrosa, Juan M. Zolezzi, Sussy Bastías-Candia, Carmen Maldonado-Bernal, David Sánchez-Herrera, M. Magdalena Aguirre-García, Alma R. Escalona-Montaño, Araceli Rojas-Bernabé, A. Pamela Gómez-García, Chandrani Fouzder, Alpana Mukhuty, Snehasis Das, Dipanjan Chattopadhyay, Amene Saghazadeh, Nima Rezaei

First published in London, United Kingdom, 2020 by IntechOpen
IntechOpen is the global imprint of INTECHOPEN LIMITED, registered in England and Wales, registration number: 11086078, 7th floor, 10 Lower Thames Street, London, EC3R 6AF, United Kingdom
Printed in Croatia

British Library Cataloguing-in-Publication Data
A catalogue record for this book is available from the British Library

Additional hard and PDF copies can be obtained from orders@intechopen.com

Toll-like Receptors
Edited by Nima Rezaei
p. cm.
Print ISBN 978-1-78984-523-5
Online ISBN 978-1-78985-391-9
eBook (PDF) ISBN 978-1-78985-392-6

For EU product safety concerns:
IN TECH d.o.o., Prolaz Marije Krucifikse Kozulić 3, 51000 Rijeka, Croatia, info@intechopen.com or visit our website at intechopen.com.

Meet the editor

Professor Nima Rezaei gained his medical degree from Tehran University of Medical Sciences and subsequently obtained an MSc in Molecular and Genetic Medicine and a PhD in Clinical Immunology and Human Genetics from the University of Sheffield, UK. He also spent a short-term fellowship studying pediatric clinical immunology and bone marrow transplantation at Newcastle General Hospital. Professor Rezaei is now Full Professor of Immunology and Vice Dean of International Affairs, School of Medicine, Tehran University of Medical Sciences, and the co-founder and Deputy President of the Research Center for Immunodeficiencies. He is also the founding president of Universal Scientific Education and Research Network. Professor Rezaei has been the director of more than 50 research projects and has designed and participated in several international collaborative projects. He is an editorial assistant or board member for more than 30 international journals. He has edited more than 10 international books, presented more than 400 lectures/posters in congresses/meetings, and published more than 700 articles in international scientific journals.

Contents

Preface

When a foreign antigen plans to reside in the body, innate immunity attempts to define the frontier lines of defense and, if necessary, helps to formulate a more specific line of defense. There is a recognition system required for innate immunity to work efficiently in such a highly responsible position. Toll-like receptors (TLRs) are an essential part of this system. They are present on innate and adaptive immune cells making them key switches incorporated in bridging innate and adaptive immunity. The non-immune cells express TLRs and, in turn, their value increases.

This book first provides readers with basic facts about TLR structure, cellular distribution, and signaling pathways. It then discusses in detail the role of natural killer cells expressing TLRs in linking innate and adaptive immunity. Furthermore, the book includes chapters that focus on the role of TLRs in infections and neurodegenerative diseases.

Nima Rezaei
Professor of Clinical Immunology,
Tehran University of Medical Sciences,
Tehran, Iran

Chapter 1

Introductory Chapter: Toll-Like Receptors

Amene Saghazadeh and Nima Rezaei

1. The first line of defense is filled with a variety of pattern recognition receptors

Pattern recognition receptors (PRRs), which are germ line-encoded receptors, probably provided the host with the best possible innate property to identify "nonself" invaders from both exogenous and endogenous sources. PRRs can discriminate self-microorganisms and molecules from nonself ones through recognizing conserved parts of microorganisms—which are known as microorganism-associated molecular patterns (MAMPs). If it is nonself, then they will direct the induction of inflammatory responses. In addition, PRRs allow the innate immunity to identify endogenous danger signals—which are released by stressed, damaged, or dying cells and known as damage-associated molecular patterns (DAMPs)—and thereby help in initiation of sterile inflammation. In this manner, PRRs participate in the clearance of invading pathogens by regulating infectious inflammation and contribute to tissue repair and regeneration in addition to elimination of autoimmunity and tumorigenesis by regulating sterile inflammatory processes.

They display three types of localization. Toll-like receptors (TLRs) are a kind of PRRs located in the membrane along with C-type lectin receptors. Also, nucleotide oligomerization domain (NOD)-like receptors (NLR) and retinoic acid-inducible gene I (RIG-I)-like receptors (RLR) are found in the nucleus and contribute to recognition of intracellular microorganisms. Finally, there are some proteins that their synthesis occurs in the cells and can be used as receptors by various microorganisms. TLRs—which are the subject of this book—were the first discovered family of PRRs.

2. Toll-like receptors

They exist in mammals and insects and mediate actions essential to achieve control over immune homeostasis. The major cells expressing TLRs are antigen-presenting cells (APCs). Activation of TLRs in APCs can affect maturation of these cells and T helper 1 (Th1) cell differentiation for processing more specific immune mechanisms [1]. However, different cell types of the body have the capacity to induce the expression of TLRs, allowing them to carry TLR-mediated signaling pathways and production of inflammatory mediators and type I interferons (IFN). When handling a number of immune and inflammatory pathways, TLRs are able to have a role in the induction of innate immune responses and to link the innate immunity with the acquired immunity. Interfering with TLR function leads inevitably to immunological anomalies seen in common conditions ranging from immunodeficiency and infection to allergy, autoimmunity, cancer and more generally to diseases of many organ systems including the central

IntechOpen

nervous system, lung, cardiovascular system, kidney, skin, and gastrointestinal system which are rooted in chronic inflammation. This has been fuel for advances in prophylactic and therapeutic applications of TLRs, especially in the last two decades.

2.1 Cells expressing TLRs

TLRs are found on the different cells of both the innate and adaptive immunity. Notably, they are expressed by nonimmune cells in the body. Normal non-transformed cells, such as endothelial cells, epithelial cells, fibroblasts, glial cells, neurons, and neural progenitor cells, as well as transformed cells of the body, i.e., cancer cells, may mediate the expression of TLRs. In humans, intestinal epithelial cells (IECs) are the main nonimmune source of TLRs. It is not aimless—there is evidence that the recognition of commensal bacteria by TLRs is required for the regulation of intestinal homeostasis [2].

2.2 Structure of TLRs

To date, 13 TLRs have been described in mammals, 10 of which are present in humans (TLR1–10). Each TLR consists of three domains: intracellular, transmembrane, and extracellular. The intracellular or cytoplasmic domain is conserved between TLRs and interleukin-1 family of receptors (IL-1R). It is, thus, referred to as the Toll-IL-1R (TIR) domain. The extracellular domain includes tandem leucine-rich repeats (LRRs), which with their curved surface appear to determine which ligand(s) a TLR can bind. The transmembrane location of TLRs makes them very suitable to transmit signals from the extracellular matrix to the cytoplasm (signal transduction).

2.3 Cellular distribution of TLRs

TLRs are located either within the cell membrane or in the intracellular compartments including the endoplasmic reticulum, endosomes, lysosomes, and endolysosomes:

Cell-membrane TLRs: TLR1, TLR2, TLR4, TLR5, TLR6, and TLR10
Intracellular TLRs: TLR3, TLR7, TLR8, and TLR9

Like other receptors, TLRs act in relation to their cellular distribution. Cell-membrane TLRs are eligible for identification of membrane components of microorganisms, e.g., lipids, proteins, and lipoproteins, while intracellular TLRs are for identification of nuclear components of microorganisms, i.e., nucleic acids.

2.4 Ligands of TLRs

In general, MAMPs are divided into three: glycans, proteins, and nucleic acids. Particularly, ligands that participate in host-pathogen interactions vary based on the pathogen identity and include:

Bacteria: lipoteichoic acid, peptidoglycan, lipoprotein/lipopeptides, deoxyribonucleic acid (DNA), flagellin, and lipopolysaccharide
Viruses: coat proteins and nucleic acids
Parasites: glycosylphosphatidylinositol (GPI) protein-membrane anchors
Yeast: zymosan

	Cellular distribution [3]	Expressing cell type				Exogenous ligands [4]			
		Innate immune cells	**Adaptive immune cells**	**IECs**	**Cancer cells**	**Bacteria Ligand (origin)**	**Viruses Ligand (origin)**	**Fungi Ligand (origin)**	**Protozoa Ligand (origin)**
TLR1	Cell surface	Monocyte/macrophage Neutrophils MDCs PDCs	B cells	×	×	Lipopeptides Soluble factors (*N. meningitidis*)	×	×	×
TLR2	Cell surface	Monocyte/macrophage Neutrophils MDCs Mast cells	×	×	Gastric cancer Colorectal cancer Ovarian cancer Cervical cancer Lung cancer Melanoma Brain cancer Breast cancer Liver cancer Laryngeal cancer Pancreatic cancer	Lipoprotein/lipopeptides (GPB, mycoplasma, mycobacteria, Spirochetes) Peptidoglycan (GPB) Lipoteichoic acid (GPB) PSM (*S. epidermidis*) HKB (*L. monocytogenes*) Porins (*Neisseria*) Soluble factors (*N. meningitidis*) Atypical LPS (*L. interrogans*, *P. gingivalis*) OmpA (*K. pneumoniae*) Glycolipids (*T. maltophilum*) LAM (mycobacteria)	HA (MV) SVP (HSV, CMV)	Zymosan (*Saccharomyces*) PLM (*C. albicans*)	GIPLs (*T. cruzi*)

	Cellular distribution [3]	Expressing cell type				Exogenous ligands [4]			
		Innate immune cells	**Adaptive immune cells**	**IECs**	**Cancer cells**	**Bacteria Ligand (origin)**	**Viruses Ligand (origin)**	**Fungi Ligand (origin)**	**Protozoa Ligand (origin)**
TLR3	Intracellular endosomal compartment	MDCs	B cells T cells	×	Colorectal cancer Ovarian cancer Cervical cancer Lung cancer Melanoma Breast cancer Liver cancer Laryngeal cancer	×	dsRNA	×	×

DOI: http://dx.doi.org/10.5772/intechopen.88493

	Cellular distribution [3]	Expressing cell type				Exogenous ligands [4]			
		Innate immune cells	Adaptive immune cells	IECs	Cancer cells	Bacteria Ligand (origin)	Viruses Ligand (origin)	Fungi Ligand (origin)	Protozoa Ligand (origin)
TLR4	Cell surface	Monocyte/macrophage Neutrophils MDCs Mast cells	B cells	✓	Gastric cancer Colorectal cancer Ovarian cancer Cervical cancer Lung cancer Prostate cancer Melanoma Brain cancer Breast cancer Liver cancer Laryngeal cancer	LPS (GNB) Hsp60 (*C. pneumonia*)	Envelope proteins (RSV and MMTV) Fusion protein (RSV)	×	GIPLs (*T. cruzi*)
TLR5	Cell surface	Monocyte/macrophage Neutrophils MDCs	×	✓	Breast cancer Gastric cancer Colorectal cancer Ovarian cancer Cervical cancer	Flagellin (flagellated bacteria)	×	×	×

	Cellular distribution [3]	**Expressing cell type**				**Exogenous ligands** [4]			
		Innate immune cells	**Adaptive immune cells**	**IECs**	**Cancer cells**	**Bacteria Ligand (origin)**	**Viruses Ligand (origin)**	**Fungi Ligand (origin)**	**Protozoa Ligand (origin)**
TLR6	Cell surface	Monocyte/macrophage Neutrophils Mast cells MDCs PDCs	B cells	×	Liver cancer	Diacyl lipopeptides (mycoplasma) Lipoteichoic acid (GPB) PSM (*S. epidermidis*) HLSF (group B streptococcus)		Zymosan (*Saccharomyces*)	
TLR7	Intracellular endosomal compartment	Monocyte/macrophage Neutrophils DCs	B cells	×	×	×	ssRNA	×	×
TLR8	Intracellular endosomal compartment	Monocyte/macrophage Neutrophils MDCs Mast cells	×	✓	×	×	ssRNA	×	×
TLR9	Intracellular endosomal compartment	Monocyte/macrophage Neutrophils PDCs	B cells T cells	×	Gastric cancer Colorectal cancer Cervical cancer Lung cancer Prostate cancer Breast cancer Liver cancer	Unmethylated CpG DNA	Unmethylated CpG DNA	×	Hemozoin (*Plasmodium*)

DOI: http://dx.doi.org/10.5772/intechopen.88493

	Cellular distribution [3]	Expressing cell type				Exogenous ligands [4]			
		Innate immune cells	Adaptive immune cells	IECs	Cancer cells	Bacteria Ligand (origin)	Viruses Ligand (origin)	Fungi Ligand (origin)	Protozoa Ligand (origin)
TLR10	Cell surface	Monocyte/macrophage	B cells	✓	×	Triacylated lipopeptides	×	×	×

GPB, gram-positive bacteria; IEC, intestinal epithelial cells; LPS, lipopolysaccharides; OmpA, outer membrane protein A; HSV, herpes simplex virus; CMV, cytomegalovirus; GIPLs, glycoinositolphospholipids; dsRNA, double-stranded RNA; GNB, gram-negative bacteria; DCs, dendritic cells; MMTV, mouse mammary tumor virus; RSV, respiratory syncytial virus; ssRNA, single-stranded RNA; MDS, myeloid dendritic cells; N. meningitides, Neisseria meningitides; PDCs, plasmacytoid dendritic cells; S. epidermidis, Staphylococcus epidermidis; PSM, phenol-soluble modulin; HKB, heat-killed bacteria; L. monocytogenes, Listeria monocytogenes; L. interrogans, Leptospira interrogans; P. gingivalis, Porphyromonas gingivalis; K. pneumoniae, Klebsiella pneumoniae; T. maltophilum, Treponema maltophilum; LAM, lipoarabinomannan; PLM, phospholipomannan; T. cruzi, Trypanosoma cruzi; C. albicans, Candida albicans; HA, hemagglutinin; MV, measles virus; SVP, structural viral proteins; C. pneumonia, Chlamydia pneumonia; HLSF, heat-labile soluble factor.

Table 1.
Toll-like receptors: cellular distribution, expressing cell types, and exogenous ligands.

TLRs have been reported to bind endogenous and exogenous ligands. There is a wide range of microorganisms including bacteria, viruses, protozoa and helminth parasites, and fungi that TLRs can defend against. **Table 1** provides an overview of ligands and associated pathogens that interact with TLRs. However, below are representative examples for MAMPs that can be recognized by TLRs:

TLR1: lipoproteins
TLR2: lipoteichoic acid, peptidoglycan, lipoproteins, and zymosan
TLR3: viral dsRNA
TLR4: lipopolysaccharide, viral envelope protein, and viral fusion protein
TLR5: flagellin
TLR6: lipoteichoic acid, lipoproteins, and zymosan
TLR7 and TLR8: viral ssRNA, synthetic antiviral compounds (imidazoquinolines)
TLR9: bacterial and viral DNA
TLR10: triacylated lipopeptides

Endogenous ligands that can be recognized by TLRs are numerous but mainly include extracellular matrix components, high-mobility group box 1, heat shock proteins (HSP), tenascin-C, cardiac myosin, and S100 proteins (for review see [5]).

3. TLR-mediated signaling pathways: network of adaptor molecules and transcription factors

Ligand recognition by TLRs involves the recruitment of different TIR domain-containing adaptor proteins. The types of adaptor proteins they use at least in part explain distinct functions of TLRs.

To date, there have been four adaptor proteins found to interact with specific TLRs (**Table 2**). Myeloid differentiation primary response 88 (MyD88) was the first of its kind. It is an intracytoplasmic adaptor molecule that consists of a C-terminal TIR domain and an N-terminal death domain. Its TIR domain—which is fundamental to ligand site recognition—can interact with all TLRs with the exception of TLR3. Recruitment of MyD88 by TLRs initiates the cascades of mitogen-activated protein kinases (MAPKs) and nuclear factor-κB (NF-κB) that result in the production of inflammatory cytokines.

In this context, TIR-domain-containing adapter-inducing interferon-β (TRIF) is another adaptor protein. TLR3 and also TLR4 can act as a trigger for TRIF-dependent signaling. This signaling then utilizes transcription factors NF-κB and interferon regulatory factor 3 (IRF3) to produce inflammatory cytokines and type I IFN.

Toll-interleukin 1 receptor (TIR) domain-containing adaptor protein (TIRAP) and TRIF-related adaptor molecule (TRAM) are last two remaining TIR domain-containing adaptor proteins. They are specialized for directing other adaptors to specific TLRs and thus are referred to as sorting adaptors. TIRAP serves as the sorting adaptor for MyD88 and facilitates its recruitment to TLR1, TLR2, TLR4, and TLR6, while TRAM is merely involved in the interaction between TRIF and TLR4.

In this manner, the discussion on signaling pathways mediated by TLRs is commonly held in MyD88-dependent and TRIF-dependent settings (for review see [7]). **Figure 1** is a schematic diagram of TLR location and pathways.

	Signaling adaptor(s)			Transcription factor(s)	Products	
	MyD88-dependent signaling	TRIF-dependent signaling	Sorting adaptor		Inflammatory cytokines	Type I IFN
TLR1	✓	×	TIRAP	NF-κB	✓	×
TLR2	✓	×	TIRAP	NF-κB	✓	×
TLR3	×	✓	×	NF-κB, IRF3, and IRF7	✓	✓
TLR4	✓	✓	TIRAP and TRAM	NF-κB, IRF3, and IRF7	✓	✓
TLR5	✓	×	×	NF-κB	✓	×
TLR6	✓	×	TIRAP	NF-κB	✓	×
TLR7	✓	×	×	NF-κB and IRF7	✓	✓
TLR8	✓	×	×	NF-κB and IRF7	✓	✓
TLR9	✓	×	×	NF-κB and IRF7	✓	✓

Table 2.
Toll-like receptor: signaling pathways and products [6].

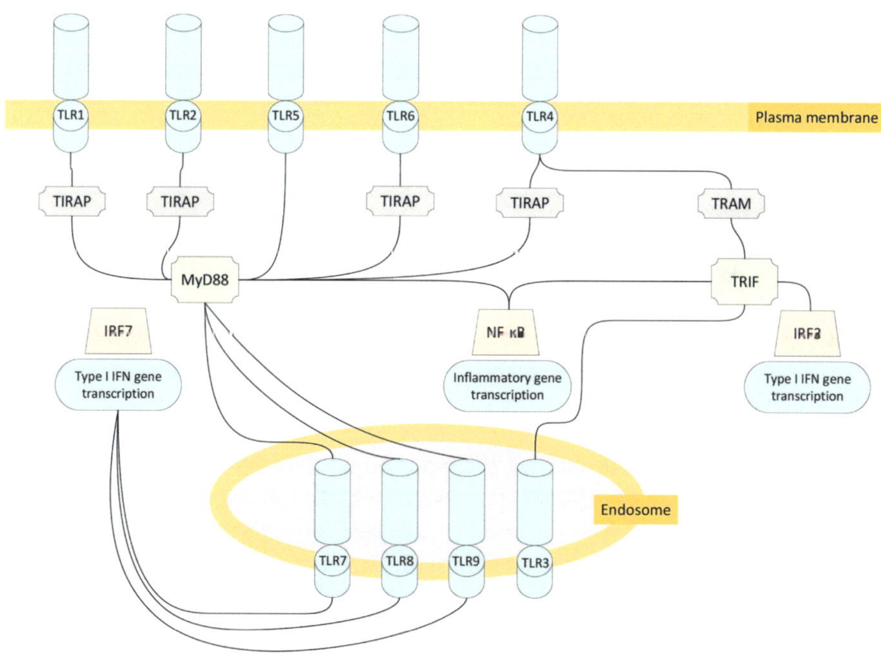

Figure 1.
Toll-like receptor (TLR) location and signaling pathways.

3.1 TLRs and different disease entities

Studies in mice have demonstrated that TLR signaling change underlies a range of pathologies. Supporting this, human studies propose that single nucleotide polymorphisms (SNPs) that alter—whether upregulated or downregulated—TLR signaling may predispose to or protect from development of diseases [8].

3.2 Autoimmunity

All 10 TLRs that are present in humans (except to TLR5) have been associated with autoimmune and inflammatory diseases including arthritis, systemic lupus erythematosus, scleroderma, and Sjogren's syndrome [9]. There are potential opposing views on involvement of TLRs in autoimmunity.

Autoimmunity is referred to as conditions where the immune system is fighting against the body itself by production of antibodies, the so-called autoantibodies, against self molecules, the so-called autoantigens. Autoantibodies bind to autoantigens and form immune complexes. The cytokine interferon-alpha (IFNα) was first thought to have a pure white role of antiviral immunity. However, the observations of autoimmune features caused by the use of recombinant IFNα in patients with chronic viral infections have expanded the former function of IFNα far beyond antiviral immunity to autoantibody production and autoimmunity. A hypothesis is that immune complexes having nucleic acids can act as ligands for TLRs, thereby making the innate immune cells to induce more than wanted or unwanted responses. TLRs that can recognize nucleic acids, i.e., TLR7 and TLR9, and plasmacytoid dendritic cells (pDCs) that express these receptors and produce IFNα in response are of particular importance in this context [9]. In contrast, some TLRs have been reported to turn the knob in the opposite direction. When reviewing the role of TLRs in inflammatory arthritis, TLRs may stimulate osteoclastogenesis, and on the other side, there are TLRs that inhibit activation of osteoclasts and thus can prevent bone destruction [10].

3.3 Brain diseases

Different brain cells reveal the expression of TLRs:

Microglia express all TLRs.
Neurons express TLR3, TLR7, TLR8, and TLR9.
Astrocytes express TLR2, TLR3, and TLR9.
Oligodendrocytes express TLR2 and TLR3.

The role of TLRs has been characterized in the normal central nervous system (CNS) as well as in disease states of the CNS. Experimental evidence suggests the possible role of TLR2 in neurogenesis, whereas TLR3 and TLR4 apparently act as downregulators of neurogenesis. Enhancement of hippocampal-dependent working memory in mice lacking TLR3 implicates this receptor as a negative regulator of cognitive functions as well. In bacterial infections of the brain and abscess formation, TLR2, TLR4, and TLR9 are essential to elicit immune responses. In the cases of viral meningitis, TLR3 and TLR9 engagement can help to localize infection and diminish neural injury as well. In parasite infections of the brain, TLR1, TLR2, and TLR9 show paradoxical effects—they may worsen disease rather than clear parasites from the brain. Both models of neuronal injury and of spinal cord injury indicate a role for TLR2 and TLR4 in inducing neuronal death and axon and myelin damage. Finally, evidence points to the potential role that TLR2, TLR4, TLR5, TLR7, and TLR9 can play in preventing the accumulation of amyloid plaques and progression of Alzheimer's disease [11].

3.4 Cardiovascular diseases

Cardiac myocytes show the expression of TLR2, TLR3, TLR4, and TLR6. TLRs play paradoxical roles in different myocardial diseases. For example, TLR2 was shown to mediate apoptosis of cardiac myocytes induced by hydrogen peroxide and doxorubicin, while TLR4 attenuated apoptosis of cardiac myocytes. Targeting both TLR2 and TLR4 provided protection in septic cardiomyopathy. TLR4 blockade implied benefit to ischemia-reperfusion injury and cardiac hypertrophy as well (for review see [12]).

TLR4 is also said to be highly expressed in atherosclerotic lesions which inflammation is supposed to corporate in its nature. There are possible explanations which can be given to this fact. The oxidization of lipids as a way to form atherosclerotic lesions accompanies thermal stress through HSP production. Oxidized lipids and HSPs can act as ligands and upregulate MyD88-dependent TLR4 relevant to inflammatory cytokine production. Another explanation is that TLR4 mediates recognition of *Chlamydia pneumoniae*, which in turn is closely related to atherosclerosis. In this manner, it would be understandable that individuals carrying Asp299Gly and Thr399Ile—which interfere with TLR4 function—develop less atherosclerotic vascular events, such as carotid stenosis, acute coronary events, acute myocardial infarction, diabetic neuropathy, and allograft rejection [13]. On the other hand, TLRs, in particular TLR2, TLR4, TLR7, and TLR9, by the aid of adenosine, can succeed in angiogenesis after myocardial injury [12].

3.5 Infections

As described above, TLR4 is critical in recognizing LPS of gram-negative bacteria (GNB). People's reactions are different to LPS inhalation and range from tolerance, i.e., no reaction, to strong asthma-like reactions. SNPs of human TLR4 gene, i.e., Asp299Gly and Thr399Ile, have been reported to affect the degree of reaction to LPS among healthy subjects and allergic asthmatic patients, development of septic shock by GNB, incidence of severe respiratory syncytial virus (RSV) bronchiolitis, risk of GNB colonization and of premature birth in pregnant women, and incidence of infections by GNB in patients on an intensive care unit [13].

3.6 Kidney diseases

Less is understood about the role of TLRs in kidney diseases. However, experimental evidence suggests that all TLRs are involved in sepsis and renal infections. Each TLR has its own associations with distinct renal diseases as well (for review see [14]).

3.7 Liver diseases

In the liver, TLR expression is observed on a variety of cells including Kupffer cells (TLR2, TLR3, TLR4, and TLR9), hepatocytes (all TLRs), hepatic stellate cells (TLR2, TLR4, and TLR9), biliary epithelial cells (TLR2, TLR3, TLR4, and TLR5), sinusoidal epithelial cells (TLR4), hepatic dendritic cells (TLR2, TLR4, TLR7, and TLR9), hepatic natural killer cells (TLR1, TLR2, TLR3, TLR4, TLR6, TLR7, and TLR9), and hepatic B cells (TLR2, TLR4, TLR7, and TLR9). Undoubtedly, such widely distrusted TLRs have been an important part of multiple liver diseases including infections of the liver by *L. monocytogenes*, and *S. typhimurium*, *P. falciparum*, hepatitis C virus, and hepatitis B virus, alcohol-induced liver diseases, nonalcoholic fatty liver disease, hepatic fibrosis, liver injury, liver regeneration, and hepatocellular carcinoma, and hepatic immune disorders (for review see [15]).

3.8 Malignancies

As for other sections, TLRs present positive and negative effects in tumorigenesis which have been discussed in [16].

3.9 Other diseases

The possible role of TLRs in periodontal health, lung diseases, and dermatological diseases has been reviewed in detail elsewhere [17–19].

4. Clinical implications

4.1 TLR agonists

TLR agonists have offered to help boosting the immune responses to vaccination as well as potential for immunotherapy of cancer, allergy, and infections. Three of which have received approval by the Food and Drug Administration (FDA) and are listed here:

Bacillus Calmette-Guérin (BCG) can act as an agonist of TLR2/TLR4 and be used for treatment of superficial transitional cell carcinoma of the bladder.
Monophosphoryl lipid A (MPL) can cause activation of TLR2/TLR4 recommended for the prophylaxis of human papilloma virus (HPV)-associated cervical cancer.
Imiquimod functions as a TLR7 agonist with implications for the treatment of actinic keratosis, basal cell carcinoma, and genital and perianal warts [20].

4.2 TLR antagonists

Antagonists targeting TLR-mediated signaling have displayed anti-inflammatory features that may be effective against invading pathogens and autoimmune diseases.

Author details

Amene Saghazadeh[1,2] and Nima Rezaei[1,3,4]*

1 Research Center for Immunodeficiencies, Children's Medical Center, Tehran University of Medical Sciences (TUMS), Tehran, Iran

2 Systematic Review and Meta-analysis Expert Group (SRMEG), Universal Scientific Education and Research Network (USERN), Tehran, Iran

3 Department of Immunology, School of Medicine, Tehran University of Medical Sciences (TUMS), Tehran, Iran

4 Network of Immunity in Infection, Malignancy and Autoimmunity (NIIMA), Universal Scientific Education and Research Network (USERN), Tehran, Iran

*Address all correspondence to: rezaei_nima@yahoo.com

References

[1] Medzhitov R. Toll-like receptors and innate immunity. Nature Reviews Immunology. 2001;**1**(2):135

[2] Rakoff-Nahoum S et al. Recognition of commensal microflora by toll-like receptors is required for intestinal homeostasis. Cell. 2004;**118**(2):229-241

[3] Pandey RK, Yu F-s, Kumar A. Targeting toll-like receptor signaling as a novel approach to prevent ocular infectious diseases. The Indian Journal of Medical Research. 2013;**138**(5):609

[4] Hedayat M, Takeda K, Rezaei N. Prophylactic and therapeutic implications of toll-like receptor ligands. Medicinal Research Reviews. 2012;**32**(2):294-325

[5] Yu L, Wang L, Chen S. Endogenous toll-like receptor ligands and their biological significance. Journal of Cellular and Molecular Medicine. 2010;**14**(11):2592-2603

[6] Kumar H, Kawai T, Akira S. Toll-like receptors and innate immunity. Biochemical and Biophysical Research Communications. 2009;**388**(4):621-625

[7] Kawai T, Akira S. The role of pattern-recognition receptors in innate immunity: Update on toll-like receptors. Nature Immunology. 2010;**11**(5):373

[8] Cook DN, Pisetsky DS, Schwartz DA. Toll-like receptors in the pathogenesis of human disease. Nature Immunology. 2004;**5**(10):975-979

[9] Marshak-Rothstein A. Toll-like receptors in systemic autoimmune disease. Nature Reviews Immunology. 2006;**6**(11):823

[10] Jiménez-Dalmaroni MJ, Gerswhin ME, Adamopoulos IE. The critical role of toll-like receptors—From microbial recognition to autoimmunity: A comprehensive review. Autoimmunity Reviews. 2016;**15**(1):1-8

[11] Hanke ML, Kielian T. Toll-like receptors in health and disease in the brain: Mechanisms and therapeutic potential. Clinical Science. 2011;**121**(9):367

[12] Frantz S, Ertl G, Bauersachs J. Mechanisms of disease: Toll-like receptors in cardiovascular disease. Nature Reviews Cardiology. 2007;**4**(8):444

[13] Schröder NWJ, Schumann RR. Single nucleotide polymorphisms of toll-like receptors and susceptibility to infectious disease. The Lancet Infectious Diseases. 2005;**5**(3):156-164

[14] Anders H-J, Banas B, Schlöndorff D. Signaling danger: Toll-like receptors and their potential roles in kidney disease. Journal of the American Society of Nephrology. 2004;**15**(4):854

[15] Seki E, Brenner DA. Toll-like receptors and adaptor molecules in liver disease: Update. Hepatology. 2008;**48**(1):322-335

[16] Rakoff-Nahoum S, Medzhitov R. Toll-like receptors and cancer. Nature Reviews Cancer. 2009;**9**(1):57

[17] Mahanonda R, Pichyangkul S. Toll-like receptors and their role in periodontal health and disease. Periodontology 2000. 2007;**43**(1):41-55

[18] McInturff JE, Modlin RL, Kim J. The role of toll-like receptors in the pathogenesis and treatment of dermatological disease. Journal of Investigative Dermatology. 2005;**125**(1):1-8

[19] Basu S, Fenton MJ. Toll-like receptors: Function and roles in lung disease. American Journal of Physiology.

Lung Cellular and Molecular Physiology. 2004;**286**(5):L887-L892

[20] Vacchelli E et al. Trial watch: FDA-approved toll-like receptor agonists for cancer therapy. Oncoimmunology. 2012;**1**(6):894-907

Chapter 2

Toll-Like Receptors and Natural Killer Cells

Carmen Maldonado-Bernal and David Sánchez-Herrera

Abstract

Natural killer (NK) cells represent a heterogeneous subpopulation of lymphocytes of the innate immune system with a powerful antitumor activity, a function given by a complex collection of receptors. They act synergistically to recognize, regulate, or amplify the response according to the microenvironment, thus highlighting Toll-like receptors (TLRs), a type of receptors that allows sensing evolutionarily molecules conserved of pathogens known as pathogen-associated molecular patterns (PAMPs) and/or damage-associated molecular patterns (DAMPs). Those TLRs are essential to start the immune response. There is little information about the different subpopulations that form NK cells as well as their expression profile of innate immune response receptors in hematological cancers.

Keywords: Toll-like receptors, natural killer cells, innate immunity, pathogen-associated molecular patterns, damage-associated molecular patterns

1. Introduction

Natural killer (NK) cells represent a highly specialized subpopulation of lymphocytes that are part of the innate immune system, whose functions vary according to the microenvironment. NK cells are involved in the early defense against foreign cells or own cells subjected to some stress (bacterial infection, viral or tumor transformation) through the recruitment of neutrophils, macrophages, dendritic cells, or B/T lymphocytes. They induce an effective adaptive response; regulate, directly or indirectly, the activity of antigen-presenting cells (APCs); and activate T lymphocytes through the natural cytotoxic activity that characterizes them or through the production of cytokines and chemokines that generate an inflammatory environment [1, 2].

NK cells play an important role in the surveillance and suppression of tumor cells; despite the significant advances that have been reached in the last decades, it is still unknown if there is a direct relationship among the population dynamics, functionality, and the phenotype of these cells. Its role in the establishment and development of malignant hematological disorders such as acute lymphoblastic leukemia (ALL), a disease characterized by the uncontrolled proliferation of B or T lymphoid precursors, is still unknown.

2. Overview of natural killer cells and toll-like receptors

2.1 NK cells

NK cells, although they are larger and present granules in their cytoplasm, morphologically are indistinguishable from the other lymphocytes. According to different authors, they comprise from 5 to 15% [3], 20% [2, 4], or even 25% [5] of total peripheral blood mononuclear cells and are derived from a CD34+ hematopoietic progenitor, as are dendritic cells (DC) and B and T lymphocytes [6].

NK cells are phenotypically defined by the expression of CD56 (neural cell adhesion molecule (NCAM) and CD16a (also known as Fcγ-RIIIA), but not CD3 and CD19, which are molecules of T and B lymphocytes, respectively [7, 8]. For a long time, they were considered as the only population of non-B or T lymphocytes. It is currently accepted that NK cells are within a subgroup of the so-called innate lymphoid cells (ILCs), whose subpopulations are differentiated according to the immunophenotype, the profile of cytokines they produce, and the transcription factors they possess [9, 10].

The NK cells were classified in group 1 of the ILCs (ILC1) due to their ability to produce INF-γ, but not cytokines such as IL-4, IL-5, IL-9, IL-13, IL-17, or IL-22, characteristic of the ILC2 and ILC3 groups, respectively [9, 11, 12]. Even within the same subgroup, they differ by having cytotoxic capacity and selectively expressing the eomesodermin (EOMES) transcription factor of other ILC non-NK that also produce INF-γ [13].

NK cells maintain a pro-inflammatory environment through the release of cytokines and recruit cells of the immune system to combat infectious agents through the release of chemokines [14] and regulate the activity of dendritic cells or activated lymphocytes [1]; they are in charge of antitumor surveillance and tolerance to healthy own cells, which conditions the rejection of transplants [15] among other functions.

In recent years, it has been reported that in addition to NK cytotoxic or regulatory NK cells, there are memory NK cells [16, 17] and NK cooperators, which secrete Th1-type cytokines (NK1) and Th2 (NK2) [18]. Even, NK cells are similar to antigen-presenting cells [19, 20], although this is in controversy.

2.2 Population diversification of NK cells and their role in the immune response

According to what was compiled by Huntington et al., NK cell precursors (NKPs) originate mainly in the bone marrow from hematopoietic precursor cells (HSCs), although they can also do so in organs such as the thymus from early lymphoid progenitor (ELP). These NKPs can mature into competent NK cells in the bone marrow or other organs, such as the liver, spleen, thymus, and lymphoid nodules, to finally enter to the circulation [21]. They migrate to other sites such as the lung, liver, mucous membranes and skin, uterus, pancreas, joints, and central nervous system (CNS), where they can exhibit unique characteristics ranging from the increase or decrease of the expression of activation receptors or effector cytotoxic molecules to the modulation of resident immune cells (**Figure 1**) [22].

According to the expression of CD56 (NCAM), NK cells can be divided into two subpopulations: NK^{Dim} and NK^{Bright}. However, according to the relative expression of the CD56 and CD16 (FcγRIIIa) markers, their functionality, and their distribution in peripheral blood or lymphoid organs, it is possible to differentiate five subpopulations of mature NK cells (**Figure 2**).

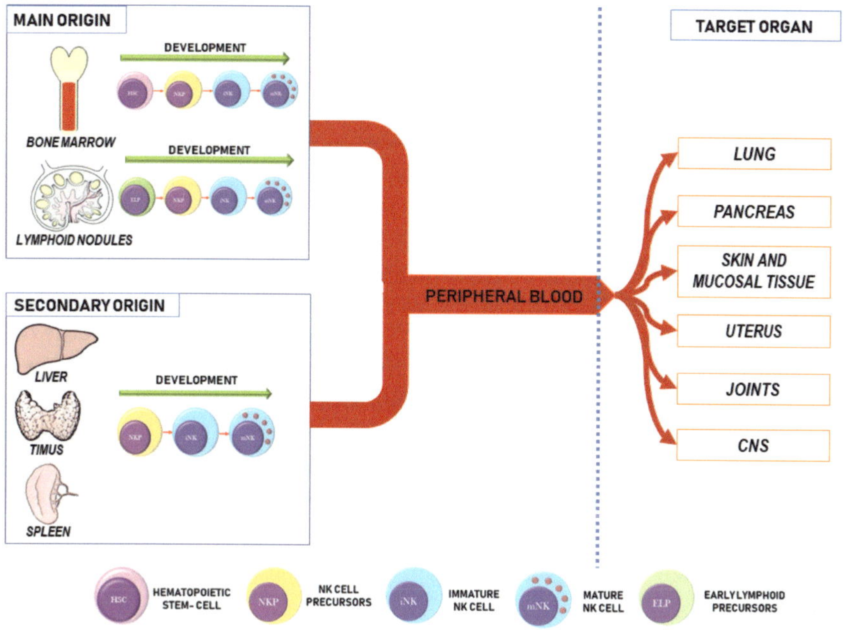

Figure 1.
Origin and distribution of NK cells in humans. The precursors of NK cell precursor (NKP) can originate from hematopoietic progenitor cells (HSC) in the bone marrow or from early lymphoid progenitor (ELP). NK cells mature mainly in the bone marrow, although the immature NKP and NK (iNK) can recirculate among the liver, spleen, and lymphoid nodes as alternative maturation sites. Mature NK cells (mNK) that leave the bone marrow reach different organs through blood circulation where they reside and modify their phenotypic and functional characteristics.

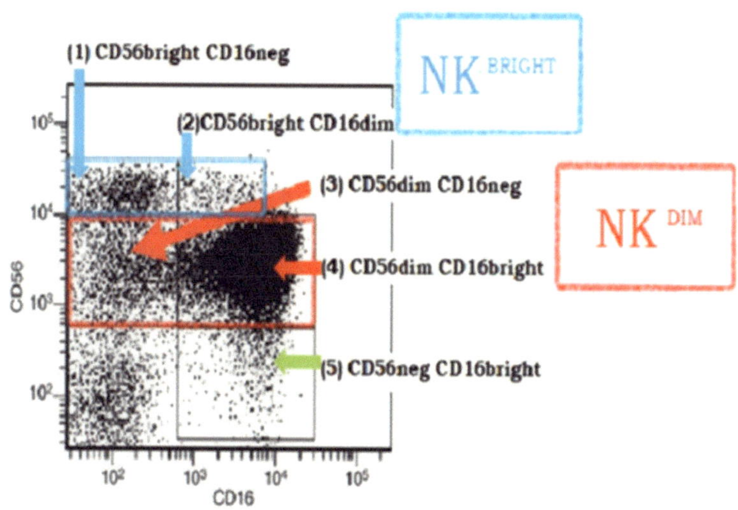

Figure 2.
Subpopulations of NK cells in peripheral blood based on the relative expression of CD56 and CD16. (1) CD56^{Bright} CD16^{Neg}, recognized for its immunoregulatory activity, represents between 50 and 70% of the CD56^{Bright} population; (2) CD56^{Bright} CD16^{Dim} represent between 30 and 50% of the CD56^{Bright} population; (3) CD56^{Dim} CD16^{Neg}; (4) CD56^{Dim} CD16^{Bright} is recognized for its cytotoxic activity; and (5) CD56^{Neg} CD16^{Bright} whose function is still unknown. Modified from [23].

The NK^{Bright} populations, also known as NK regulators, comprise about 10% of total peripheral blood NK cells. From these, about 50–70% have a [CD56^{High} CD16^{Neg}] phenotype and 30–50% a [CD56^{High}/CD16^{Low}] phenotype

[23]; they are mainly characterized by their poor cytotoxic capacity and their high capacity to secrete several types of post-activation cytokines, mainly INF-γ but also TNF-β, IL-5, IL-10, and IL-13 [7, 8] and constitutively some chemokines such as MIP-1α, MIP-1β, and RANTES [24]. They also have the ability to proliferate in culture when exposed to low doses of IL-2 (picomoles) compared to the cytotoxic NK cells, which does not show an evident proliferation under the same conditions [25]. The NK^{Bright} cells [$CD56^{High}$ $CD16^{Neg}$] are assigned an exclusively regulatory function, since their cytotoxic activity is poor and their genetic profile is directed mainly to the production of cytokines and not to the cytotoxic activity, in comparison with the NK^{Dim}, whereas the NK^{Bright} subpopulation [$CD56^{High}$ $CD16^{Low}$] is considered more as a transition phenotype [26]; because despite having the same characteristics of the previous phenotype, it has a lower rate of cell division when stimulated with IL-2 and does not change its activity even with ligands for c-KIT [27]. It exhibits cytotoxic activity [28], and it has also been seen that it represents the highest percentage of NK cells in circulation in bone marrow transplants, until its normalization around the fourth month [29, 30]. NK^{Bright} cells are usually not found in peripheral blood, bone marrow, and spleen as they are mainly distributed in secondary lymphoid nodules (parafollicular zone of T cells) [31] and tonsils [32].

On the other hand, the NK^{Dim} population represents around 90% of peripheral blood NK cells; they have a phenotype [$CD56^{Low}$ $CD16^{Neg}$] whose main function has not been well established [23] and a main phenotype [$CD56^{Low}$ $CD16^{High}$] that exhibits potent cytotoxic activity. Although they are generally poor producers of cytokines [7, 8, 33], they tend to predominate in the spleen, peripheral blood, and bone marrow [32, 34].

It is important to clarify that the behavior of each subpopulation, in terms of post-activation secretion of cytokines and chemokines, will depend largely on the stimuli they receive either by recognizing target cells (tumor or transformed) or responding to exogenous cytokines.

Compared with NK^{Dim} cells, NK^{Bright} cells produce more TNF-α and INF-γ after activation with PMA/ionomycin [35] or exogenous cytokines such as IL-12, IL-15, or IL-18, alone or in combination [8].

On the contrary, when it comes to a response to the recognition of target cells, NK^{Dim} cells significantly increase their production of cytokines and chemokines compared to NK^{Bright}, such as MCP-1 (CCL2), IL-8 (CXCL8), IP-10 (CXCL10), soluble IL-2Rα (CD25), GM-CSF, and IL-5 and low levels of IL-1β, IL-6, IL-7, IL-10, IL-12p40, IFN-α, and MIG (CXCL9). In addition, it increases the production of chemokines such as MIP-1α (CCL3), MIP-1β (CCL4), and RANTES (CCL5) that produce constitutively [24].

The cytotoxic activity exhibited by NK cells does not require prior sensitization to kill their target cells, since it is not dependent on the presentation of a specific antigen as in the case of $CD8^+$ T cells [7, 36] and can be mediated through membranolitic and/or apoptotic mechanisms (**Figure 3**).

The membranolitic mechanisms include the production of perforins, enzymes that when integrated into the cell membrane form a pore that allows water to enter and cause osmotic lysis [37]. In the past, it was believed that both NK^{Bright} and NK^{Dim} cells had similar levels of perforins [38]; however, more recent studies by flow cytometry indicate that NK^{Dim} cells have at least 10 times more perforins than NK^{Bright} ones [35].

Regarding apoptotic mechanisms, these can induce the death of the target cell through complex mechanisms that involve both death-inducing proteins and specific ligand-receptor interactions through one of the following routes:

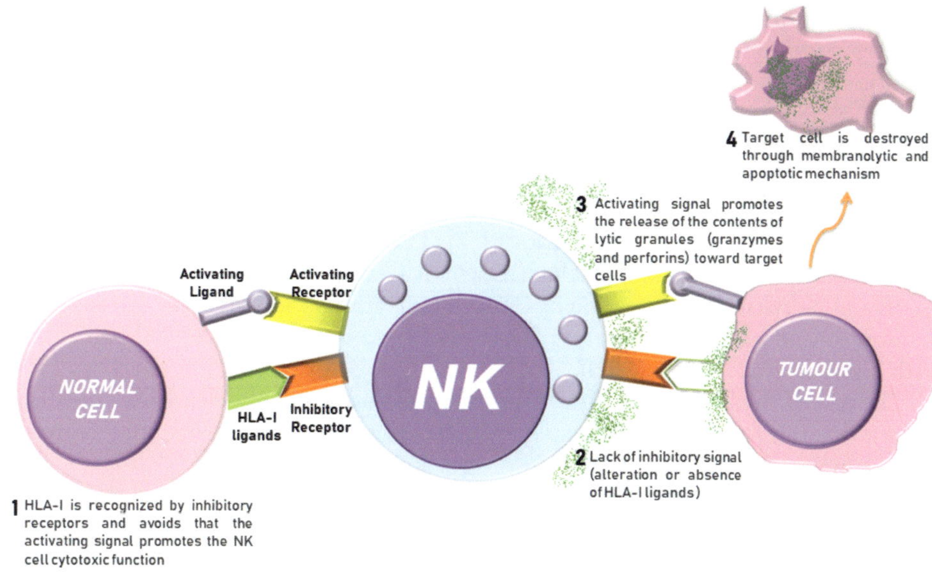

Figure 3.
Recognition and elimination of abnormal cells by NK cells. NK cells possess the ability to discriminate normal cells of tumor or transformed cells by detecting alterations at the HLA-I level; the target cells are eliminated by membranolitic and/or apoptotic mechanisms.

2.3 Granzyme pathway

Granzymes are proteins capable of activating the apoptosis program [39] following two mechanisms. The first does not depend on the activity of caspase proteins and is mediated mainly by granzyme A, and this fraction is single-stranded DNA (ssDNA) and interferes in the repair of genetic material without producing cell lysis [40, 41], while the second promotes the activity of caspase proteins and is mediated mainly by granzyme B [42].

The NK cell presents granzymes A, B, K, and M; NK^{Dim} cells present a high expression of granzymes A and B, whereas NK^{Bright} cells mainly express granzyme K [35, 43]. There are reports that NK cells express almost exclusively granzyme M; this enzyme is capable of mediating cell death independent of the activation of caspase proteins and in the presence of perforins, without fractionating DNA or producing changes in mitochondria [44].

In mice, deficient in granzymes A/B and/or perforins, it has been seen that there is uncontrolled growth of solid tumors, which suggests that these enzymes play an important role in the immunosurveillance of tumor cells mediated by NK cells [45].

2.4 NK cell ligand binding pathway to the cell death receptors expressed by the target cell

They include Fas ligands [FasL (CD95L)-Fas (CD95)] and/or the ligand that induces apoptosis related to tumor necrosis factor α (TRAIL) [46, 47].

2.5 Antibody-dependent cellular cytotoxicity (ADCC)

NK cells express FcγRIIC/CD32c [48] and FcγRIIIA/CD16a [34]. These receptors interact with opsonized target cells, through the Fc regions of the antibodies, which combined with cellular antigens that cause the death of the target cell [49]. To through of mechanisms that involve the release of cytotoxic granules (perforin-granzyme), or by stimulation of apoptosis through of TNF-related apoptosis-inducing ligand (TRAIL) and/or by release of pro-inflammatory cytokines that promote the activity of other cells [50].

NK^{Dim} cells with [$CD56^{Low}$ $CD16^{High}$] phenotype direct this mechanism in comparison with NK^{Bright} cells. Although it has been seen that the subpopulation with [$CD56^{High}$ $CD16^{Low}$] phenotype exhibits low cytotoxic activity [51], NK^{Dim} cells [$CD56^{Low}$ $CD16^{Neg}$] show a higher antitumor activity against cell lines (natural cytotoxicity) than other subpopulations [52]. This is supported by other studies where it is reported that NK^{Dim} cells [$CD56^{Low}$ $CD16^{High}$] lose the expression of CD16 and increase the expression of CD107a (a degranulation marker), through a disintegrin and a metalloprotease-17 (ADAM-17), to become [$CD56^{Low}$ $CD16^{Neg}$] with high cytotoxic capacity [53].

The role of the [$CD56^{Neg}$ $CD16^{High}$] subpopulation is still not clearly defined. It is known that it is found in a low frequency in healthy individuals. It does not express surface molecules of other lymphoid lineages and that in chronic viral diseases, such as the human immunodeficiency virus (HIV). It presents changes in the level of expression of their activity receptors, characterized by the increase in the expression of inhibitory receptors and the decrease of natural cytotoxicity receptors (NCRs), together with other effector molecules that are hardly observed in healthy people [54–56].

It is considered that the [$CD56^{Neg}$ $CD16^{High}$] subpopulation is dysfunctional in terms of its lytic and antiviral activity, although it retains the ability to produce pro-inflammatory chemokines [54–56].

Zulu et al. demonstrated that the HIV induces the expansion of the negative CD56 population of NK cells through the upregulation of NKG2C receptors and the negative regulation of Siglec-7, NKG2A, and CD57 receptors [57].

2.6 Receptors of the NK cells

NK cells have signals through a wide variety of receptors that allow them to respond to different types of stimuli and grant great flexibility when exercising their effector and/or cytotoxic function.

The function of the NK cell is given by a complex collection of receptors that act in a synergistic way to recognize, regulate, or amplify the response according to the microenvironment. Thus highlighting the pattern recognition receptors (PRRs), such as Toll-like receptors or natural cytotoxicity receptors, and inhibitory killing receptors (iNKRs), such as receptors that are activated during early response to pathogens, cells transformed by virus or tumor cells [58].

PRRs are a family of innate immune response receptors that recognize evolutionarily conserved microbial products whose activation favors the production of pro-inflammatory cytokines. Within the PRR group, the TLRs are the most studied, although they are not the only ones; there are also the NOD-like receptors (NLRs) and the retinoid acid-inducible gene I (RIG-I)-like receptors (RLRs) [59].

NK cells express innate immune response receptors, such as NOD2, NLRP3, TLR3, TLR7, and TLR9, and promote the production of inflammatory cytokines and chemokines that are capable of amplifying the immune response [60]. The modulation of these cells through their innate immune response receptors, mainly via TLR, has gained interest and represents a promising therapeutic alternative against conditions such as cancer. Since there have been studies for a long time that support the possibility of its use, it has been observed that when ODNs (ligands of TLR9) are intraperitoneally administered in lymphoma murine models, an effective elimination of tumor cells occurs in 80% of cases [61].

2.7 Toll-like receptors and their role in NK cells

Toll-like receptors are among the most important group of pattern recognition receptors, since they orchestrate a wide variety of activities related to the immune response.

These receptors recognize a wide variety of molecules evolutionarily conserved, associated with microorganisms, such as lipopolysaccharides, lipoproteins, mycolic acids, non-methylated DNA, and double-stranded RNA, generically known as pathogen-associated molecular patterns [62–64]. TLRs also recognize endogenous molecules called damage-associated molecular patterns, which originate from damaged cells [65] or are products of altered metabolism of transformed cells in conditions such as cancer [66, 67] and autoimmune diseases [67–70] or associated with chronic inflammation [71, 72]. They play an important role in the evolution of these conditions.

2.8 Overview of the toll-like receptors

Structurally, TLRs are type I integral glycoproteins that present an extracellular domain with leucine-rich repeats (LRRs) that are responsible for binding and discriminating ligands (PAMPs or DAMPS) present in the cellular microenvironment. They have a transmembrane domain and an intracellular Toll/interleukin (IL)-1 receptor (TIR) domain that triggers the signaling cascade via MyD88/TRIF and is highly conserved among each subfamily of TLRs [73].

There are 13 TLRs described in mammals, and 10 are found at the protein level in humans and differ according to their cellular localization and to the different PAMPs/DAMPs to which they respond. TLR11 in humans is a pseudogene, so it is not expressed [74].

The TLRs that are found mainly in the cell membrane are TLRs 1, 2, 4, 5, and 6. They sense structural components of bacteria, fungi, helminthes, or protozoa, whereas TLRs that are mainly found in intracellular compartments, such as TLRs 3, 7, 8, and 9, sense nucleic acids of viral and/or bacterial origin [73, 75] (**Figure 4**).

The stimulation of the TLRs is capable of initiating an immune response to various stimuli on its own, as well as of controlling the adaptive response through the inflammatory process with the production of pro-inflammatory cytokines (IL-1β, IL-6, TNF-α), chemokines (IL-8, MCP-1) [76]; defensins [77]; type I interferons [78]; co-stimulation and MHC molecules [79]. The union between the different types of immune responses through the TLRs takes as a classic example the dendritic cells. They inspect their environment through the TLRs [80]; and once they detect a ligand (bacterial product, viral or stress protein), increase the expression of co-stimulatory molecules capable of stimulating T naive cells [81] and polarizes the adaptive immune response toward Th1 or Th2 profiles [82].

The cellular response through direct stimulation with TLR ligands will depend largely on the type of lineage in question. The information that is reported about activity and expression in NK cells is relatively new, and it is more associated with innate antibacterial or antiviral immune response [83], but not in cancer.

2.9 Expression of TLRs in NK cells

The expression profile of TLRs in NK cells was initially limited to the detection of mRNA. However, the results do not always reflect the expression of the protein since it is difficult to identify the receptors in the NK cell.

NK cells express most of the human TLRs reported to date, although the detection and level of mRNA expression of each receptor vary depending on the author.

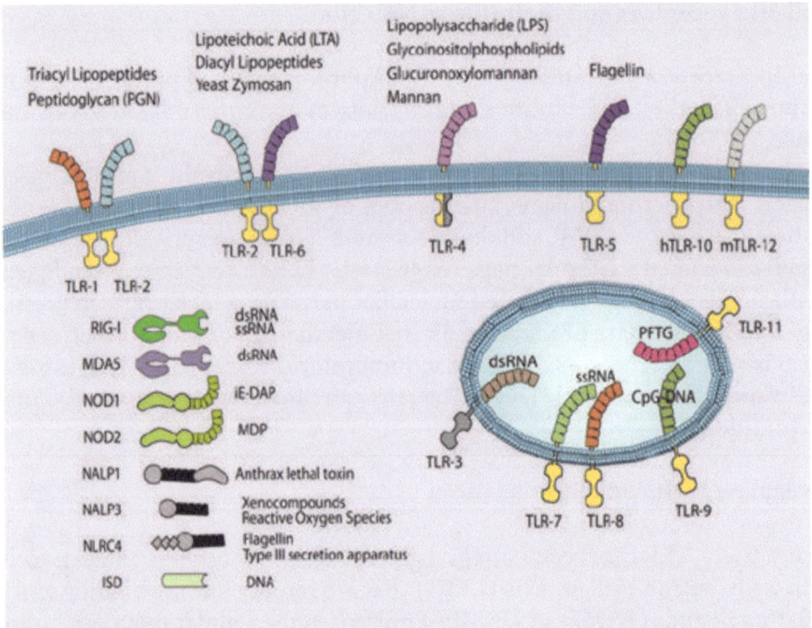

Figure 4.
Toll-like receptors and their ligands. TLRs are transmembrane proteins of a glycoprotein nature that possess the ability to sense highly conserved microorganism molecules known as pathogen-associated molecular patterns (PAMPs), such as flagellin, LPS, or genetic material (ssDNA, ssRNA, dsDNA, dsRNA). In humans, 10 of the 13 TLRs present in mammals have been detected [84].

There are authors who indicate that NK cells express high levels of TLR1, followed by TLR2, TLR3, TLR5, TLR6, and low levels of TLR4 [85, 86]. Other authors agree that TLR2 and TLR3 have greater expression, followed by TLR5 and TLR6; in those studies, TLR1 mRNA was not quantified [87–89]. On the other hand, TLR7 [85, 89, 90] and TLR10 [83, 86] present levels so low that they are practically undetectable.

There is a controversy about whether or not NK cells express TLR8 [86, 87, 89] and TLR9 [86, 88], although some authors point out that the cells constitutively express mRNA of all TLRs [85, 91, 92].

NK cells have higher levels of TLR3 mRNA than any other peripheral blood mononuclear cell, such as monocytes, B and T lymphocytes, or plasmacytoid dendritic cells [85].

Through techniques such as flow cytometry, Western blot, and immunoprecipitation, it is possible to know that NK cells of healthy people have a defined TLR expression profile (**Table 1**) and the expression of receptors is independent of their activation state [58].

In addition, there are variations in the level of TLR expression within the same subpopulations of NK cells. It is accepted that both NK^{Bright} and NK^{Dim} exhibit a similar mRNA profile of TLRs, although it is not always reflected at the protein level and there is a great controversy regarding the distribution and presence of some of these receptors in both subpopulations, especially TLR2, TLR4, and TLR3.

TLR2 and TLR4 are mainly distributed on the cell surface, whereas TLR3 is generally found in intracellular vesicles [75]; however, it has been seen that in NK cells, TLR3 is expressed both within [93] and on the cell surface [94]. There are publications reporting that TLR2 and TLR4 exhibit a marked intracellular distribution [95], although other authors indicate otherwise [96, 97].

The relative amount of some TLRs may vary according to the phenotype (**Dim** or **Bright**), although expression levels appear to be higher in cells with regulatory

TLR	mRNA[1]	Protein	Detection method
		Presence	
1	Very high	Yes	Flow cytometry [97] Western blot [97]
2	High/moderate	Yes[3]	Direct activation of the TLR2-/MyD88-dependent pathway [96] Flow cytometry [95–97] Western blot
3	High/moderate	Yes	Flow cytometry [93, 94] Western blot [94]
4	Low	Yes[4]	Flow cytometry [94, 95, 99]
5	High/moderate	Yes[5]	S/R
6	High/moderate	Yes	Flow cytometry and Western blot [97]
7	Very low/undetectable	Yes	Flow cytometry [93, 94] Western blot [94]
8	Low[2]	Yes	Flow cytometry [94] Western blot [90, 94]
9	Low[2]	Yes	Flow cytometry [93, 95, 100] Western blot [100]
10	Very low/undetectable	N/R	N/R

N/R, not reported.[1]*The levels of relative expression are given according to what was reported by [85, 91] and refer to the comparison of expression among the 10 TLRs.*
[2]*There is a controversy whether or not they express mRNA of these TLRs, since some reports indicate that it was not possible to detect it.*
[3]*In previous studies, TLR2 could not be detected by flow cytometry or by immunoprecipitation.*
[4]*More recent studies indicate that it is expressed mainly as intracellular [95, 99].*
[5]*No reports were found indicating the presence of TLR5; however it is inferred that it is present as it responds specifically to flagellin [88, 101, 102], a molecule that it is only recognized through this receptor.*

Table 1.
Expression of TLRs in human natural killer cells.

phenotype [89], which suggests that the type of response could be conditioned to promote a cytotoxic or immunomodulatory response when using one ligand or another in TLR activation assays (**Table 2**).

It has been seen that NK^{Bright} cells express more TLR1, TLR2, and TLR6 than NK^{Dim} [97], although other studies report that less than 1% of total NK cells express these three receptors [98].

NK^{Dim} cells can express, under normal conditions, more TLR4 than NK^{Bright} cells [99] although other authors seem to find no differences in the expression in TLR2, TLR4 [95], and TLR9 [95, 100].

There is no information about whether there is differential expression of TLRs 3, 5, 7, or 8, and the distribution pattern of TLR5 is not known. However, it is inferred that NK cells express it, since they respond to flagellin and there are several studies that demonstrate it [88, 101, 102]. To date there are no reports about the presence, distribution, or role of TLR10 in NK cells.

The therapeutic use of TLR ligands in the modulation of NK cells against cancer, especially in malignant hematological disorders such as leukemia, is an interesting alternative for the treatment of this type of diseases, since there are reports that reveal their therapeutic use as potential antitumor agents and as adjuvants in vaccines and other therapeutic modalities [103]. It is currently the subject of an extensive review by several research groups [104, 105].

TLR	Cellular localization	Population distribution
1	Extracellular [97]	$NK^{Bright} > NK^{Dim}$ [97]
2	Extracellular [96, 97] and intracellular [95]	$NK^{Bright} > NK^{Dim}$ [97]
3	Intracellular [93] and extracellular[1] [94]	N/R
4	Extracellular [94] and intracellular[2] [95, 99]	$NK^{Bright} < NK^{Dim}$ [99] $NK^{Bright} = NK^{Dim}$ [95]
5	N/R	N/R
6	Extracellular [97]	$NK^{Bright} > NK^{Dim}$ [97]
7	Intracellular [93, 94]	N/R
8	Intracellular [94]	N/R[4]
9	Mainly intracellular[3] [93, 95, 100]	$NK^{Bright} = NK^{Dim5}$ [95]
10	N/R	N/R

N/R, not reported.[1]It was found that both, in cell lines (NKL, NK92, and YT) and in NK of peripheral blood, TLR3 is expressed on the surface [94].
[2]Studies that are more recent indicate that it is mainly expressed as intracellular [95, 99].
[3]TLR9 expression exists in plasma membrane, but it is quite low compared to intracellular expression.
[4]There are no studies that determine whether there is differential expression, although it has been seen that NKBright cells are better activated with ssRNA40 than NKDim cells, suggesting that the latter have a lower expression of TLR8.
[5]In other studies, it seems that NKDim cells express more TLR9 than NKBright cells and its expression conditions the response to ligands of this TLR [100].

Table 2.
Localization and differential distribution of TLRs in human natural killer cells.

3. Conclusion

In this chapter, we included the overview of NK cells, their population diversification and role in the immune response, and their expression and role of TLRs.

Acknowledgment

We thank doctors Jacques Zimmer, MD, PhD, and Sankar Ghosh, PhD, for allowing us to use some of their figures, **Figures 2** and **4**, in this chapter of the book.

Conflict of interest

The authors declare that there is no conflict of interest.

Author details

Carmen Maldonado-Bernal* and David Sánchez-Herrera
Immunology and Proteomics Research Laboratory, Children's Hospital of Mexico
Federico Gómez, Mexico City, Mexico

*Address all correspondence to: cmaldobe@yahoo.com

References

[1] LongEO KHS, Liu D, Peterson ME, Rajagopalan S. Controlling NK cell responses: Integration of signals for activation and inhibition. Annual Review of Immunology. 2013;**31**:227-258. DOI: 10.1146/annurev-immunol-020711-075005

[2] Campbell KS, Hasegawa J. Natural killer cell biology: An update and future directions. The Journal of Allergy and Clinical Immunology. 2013;**132**(3):536-544. DOI: 10.1016/j.jaci.2013.07.006

[3] Cooper MA, Fehniger TA, Caligiuri MA. The biology of human natural killer-cell subsets. Trends in Immunology. 2001;**22**(11):633-640

[4] Rojas-Pandales F, Bolaños N, Mercado M, Gonzále JM, Cuéllar A, Cifuentes-Rojas C. Reference values of the natural killer cells (NK and NKT) on blood donors in Bogota. Acta médica colombiana. 2007;**32**(3):124-128

[5] Pittari G, Filippini P, Gentilcore G, Grivel JC, Rutella S. Revving up natural killer cells and cytokine-induced killer cells against hematological malignancies. Frontiers in Immunology. 2015;**6**:230. DOI: 10.3389/fimmu.2015.00230

[6] Galy A, Travis M, Cen D, Chen B. Human T, B, natural killer, and dendritic cells arise from a common bone marrow progenitor cell subset. Immunity. 1995;**3**(4):459-473

[7] Caliguri MA. Human natural killer cells. Blood. 2008;**112**(3):461-469. DOI: 10.1182/blood-2007-09-077438

[8] Cooper MA, Fehniger TA, Turner SC, Chen KS, Ghaheri BA, Ghayur T, et al. Human natural killer cells: A unique innate immunoregulatory role for the CD56 bright subset. Blood. 2001;**97**:3146-3151

[9] Walker JA, Barlow JL, McKenzie AN. Innate lymphoid cells—How did we miss them? Nature Reviews Immunology. 2013;**13**:75. DOI: 10.1038/nri3349

[10] Spits H, Artis D, Colonna M, Diefenbach A, Di Santo JP, Eberl G, et al. Innate lymphoid cells—A proposal for uniform nomenclature. Nature Reviews Immunology. 2013;**13**:145-149. DOI: 10.1038/nri3365

[11] Artis D, Spits H. The biology of innate lymphoid cells. Nature. 2015;**517**(7534):293-301. DOI: 10.1038/nature14189

[12] Hazenberg MD, Spits H. Human innate lymphoid cells. Blood. 2014;**124**(5):700-709. DOI: 10.1182/blood-2013-11-427781

[13] Gordon SM, Chaix J, Rupp LJ, Wu J, Madera S, Sun JC, et al. The transcription factors T-bet and Eomes control key checkpoints of natural killer cell maturation. Immunity. 2012;**36**(1):55-67. DOI: 10.1016/j.immuni.2011.11.016

[14] Souza-Fonseca-Guimaraes F, Adib-Conquy M, Cavaillon JM. Natural killer (NK) cells in antibacterial innate immunity: Angels or devils? Molecular Medicine. 2012;**18**(1):270-285. DOI: 10.2119/molmed.2011.00201

[15] Benichou G, Yamada Y, Aoyama A, Madsen JC. Natural killer cells in rejection and tolerance of solid organ allografts. Current Opinion in Organ Transplantation. 2011;**16**(1):47. DOI: 10.1097/MOT.0b013e32834254cf

[16] Sun JC, Beilke JN, Lanier LL. Adaptive immune features of natural killer cells. Nature. 2009;**457**(7229):557. DOI: 10.1038/nature07665

[17] O'Sullivan TE, Sun JC, Lanier LL. Natural killer cell memory. Immunity. 2015;**43**(4):634-645

[18] Kimura MY, Nakayama T. Differentiation of NK1 and NK2 cells. Critical Reviews in Immunology. 2005;**25**(5):361-374. PMID: 16167886

[19] Spits H, Lanier LL. Natural killer or dendritic: What's in a name? Immunity. 2007;**26**(1):11-16

[20] Hanna J, Mandelboim O. Novel APC-like properties of human NK cells directly regulate T cell activation. The Journal of Clinical Investigation. 2015;**125**(4):1763. DOI: 10.1172/JCI81527

[21] Huntington ND, Vosshenrich CA, Di Santo JP. Developmental pathways that generate natural-killer-cell diversity in mice and humans. Nature Reviews Immunology. 2007;**7**(9):703-714. DOI: 10.1038/nri2154

[22] Shi FD, Ljunggren HG, La Cava A, Van Kaer L. Organ-specific features of natural killer cells. Nature Reviews Immunology. 2011;**11**(10):658

[23] Poli A, Michel T, Thérésine M, Andrès E, Hentges F, Zimmer J. CD56bright natural killer (NK) cells: An important NK cell subset. Immunology. 2009;**126**(4):458-465. DOI: 10.1111/j.1365-2567.2008.03027.x

[24] Fauriat C, Long EO, Ljunggren HG, Bryceson YT. Regulation of human NK-cell cytokine and chemokine production by target cell recognition. Blood. 2010;**115**(11):2167-2176. DOI: 10.1182/blood-2009-08-238469

[25] Carson WE, Giri JG, Lindemann M, Linett ML, Ahdieh M, Paxton R, et al. Interleukin (IL) 15 is a novel cytokine that activates human natural killer cells via components of the IL-2 receptor. The Journal of Experimental Medicine. 1994;**180**(4):1395-1403

[26] Béziat V, Duffy D, Quoc SN, Le Garff-Tavernier M, Decocq J, Combadière B, et al. CD56brightCD16$^+$ NK cells: A functional intermediate stage of NK cell differentiation. Journal of Immunology. 2011;**186**(12):6753-6761. DOI: 10.4049/jimmunol.1100330

[27] Carson WE, Fehniger TA, Caligiuri MA. CD56 bright natural killer cell subsets: Characterization of distinct functional responses to interleukin-2 and the c-kit ligand. European Journal of Immunology. 1997;**27**(2):354-360. DOI: 10.1002/eji.1830270203

[28] Takahashi E, Kuranaga N, Satoh K, Habu Y, Shinomiya N, Asano T, et al. Induction of CD16+ CD56bright NK cells with antitumour cytotoxicity not only from CD16− CD56bright NK cells but also from CD16− CD56dim NK cells. Scandinavian Journal of Immunology. 2007;**65**(2):126-138. DOI: 10.1111/j.1365-3083.2006.01883.x

[29] Gottschalk LR, Bray RA, Kaizer H, Gebel HM. Two populations of CD56 (Leu-19)+/CD16+ cells in bone marrow transplant recipients. Bone Marrow Transplantation. 1990;**5**(4):259-264

[30] Jacobs R, Stoll M, Stratmann G, Leo R, Link H, Schmidt RE. CD16-CD56+ natural killer cells after bone marrow transplantation. Blood. 1992;**79**(12):3239-3244

[31] Fehniger TA, Cooper MA, Nuovo GJ, Cella M, Facchetti F, Colonna M, et al. CD56 bright natural killer cells are present in human lymph nodes and are activated by T cell–derived IL-2: A potential new link between adaptive and innate immunity. Blood. 2003;**101**(8):3052-3057. DOI: 10.1182/blood-2002-09-2876

[32] Ferlazzo G, Thomas D, Lin SL, Goodman K, Morandi B, Muller WA, et al. The abundant NK cells in human secondary lymphoid tissues require

activation to express killer cell Ig-like receptors and become cytolytic. Journal of Immunology. 2004;**172**(3):1455-1462. DOI: 10.4049/jimmunol.172.3.1455

[33] Lanier LL, Ruitenberg JJ, Phillips JH. Functional and biochemical analysis of CD16 antigen on natural killer cells and granulocytes. Journal of Immunology. 1988;**141**(10):3478-3485

[34] Yokoyama WM, Kim S, French AR. The dynamic life of natural killer cells. Annual Review of Immunology. 2004;**22**:405-429. DOI: 10.1146/annurev.immunol.22.012703.104711

[35] Jacobs R, Hintzen G, Kemper A, Beul K, Kempf S, Behrens G, et al. CD56bright cells differ in their KIR repertoire and cytotoxic features from CD56dim NK cells. European Journal of Immunology. 2001;**31**(10):3121-3126. DOI: 10.1002/1521-4141(2001010)31:10<3121:AID-IMMU3121gt;3.0.CO;2-4

[36] Kärre K. Natural killer cell recognition of missing self. Nature Immunology. 2008;**9**(5):477-480. DOI: 10.1038/ni0508-477

[37] Podack ER. Functional significance of two cytolytic pathways of cytotoxic T lymphocytes. Journal of Leukocyte Biology. 1995;**57**(4):548-552

[38] Konjevic G, Schlesinger B, Cheng L, Olsen KJ, Podack ER, Spuzic I. Analysis of perforin expression in human peripheral blood lymphocytes, CD56^{+} natural killer cell subsets and its induction by interleukin-2. Immunological Investigations. 1995;**24**(3):499-507

[39] Simon MM, Hausmann M, Tran T, Ebnet K, Tschopp J, ThaHla R, et al. In vitro- and ex vivo-derived cytolytic leukocytes from granzyme A × B double knockout mice are defective in granule-mediated apoptosis but not lysis of target cells. The Journal of Experimental Medicine. 1997;**186**(10):1781-1786

[40] Beresford PJ, Xia Z, Greenberg AH, Lieberman J. Granzyme A loading induces rapid cytolysis and a novel form of DNA damage independently of caspase activation. Immunity. 1999, 1999;**10**(5):585-595

[41] Lieberman J, Fan Z. Nuclear war: The granzyme A-bomb. Current Opinion in Immunology. 2003;**15**(5):553-559

[42] King KL, Cidlowski JA. Cell cycle regulation and apoptosis. Annual Review of Physiology. 1998;**60**(1): 601-617. DOI: 10.1146/annurev.physiol.60.1.601

[43] Bratke K, Kuepper M, Bade B, Virchow JC, Luttmann W. Differential expression of human granzymes A, B, and K in natural killer cells and during CD8+ T cell differentiation in peripheral blood. European Journal of Immunology. 2005;**35**(9):2608-2616

[44] Kelly JM, Waterhouse NJ, Cretney E, Browne KA, Ellis S, Trapani JA, et al. Granzyme M mediates a novel form of perforin-dependent cell death. The Journal of Biological Chemistry. 2004;**279**(21):22236-22242. DOI: 10.1074/jbc.M401670200

[45] Pardo J, Balkow S, Anel A, Simon MM. Granzymes are essential for natural killer cell-mediated and perf-facilitated tumor control. European Journal of Immunology. 2002;**32**(10):2881-2886. DOI: 10.1002/1521-4141(2002010)32:10<2881

[46] Zamai L, Ahmad M, Bennett IM, Azzoni L, Alnemri ES, Perussia B. Natural killer (NK) cell-mediated cytotoxicity: Differential use of TRAIL and Fas ligand by immature and mature primary human NK cells. The Journal of Experimental Medicine. 1998;**188**(12):2375-2380

[47] Srivastava S, Lundqvist A, Childs RW. Natural killer cell immunotherapy

for cancer: A new hope. Cytotherapy. 2008;**10**(8):775-783. DOI: 10.1080/14653240802648181

[48] Morel PA, Ernst LK, Metes D. Functional CD32 molecules on human NK cells. Leukemia & Lymphoma. 1999;**35**(1-2):47-56. DOI: 10.3109/10428199909145704

[49] Leibson PJ. Signal transduction during natural killer cell activation: Inside the mind of a killer. Immunity. 1997;**6**(6):655-661. DOI: org/10.1016/S1074-7613(00)80441-0

[50] Smyth MJ, Cretney E, Kelly JM, Westwood JA, Street SE, Yagita H, et al. Activation of NK cell cytotoxicity. Molecular Immunology. 2005;**42**(4):501-510. DOI: 10.1016/j.molimm.2004.07.034

[51] Nagler A, Lanier LL, Cwirla S, Phillips JH. Comparative studies of human FcRIII-positive and negative natural killer cells. Journal of Immunology. 1989;**143**(10):3183-3191

[52] Penack O, Gentilini C, Fischer L, Asemissen AM, Scheibenbogen C, Thiel E, et al. CD56 dim CD16 neg cells are responsible for natural cytotoxicity against tumor targets. Leukemia. 2005;**19**(5):835-840. DOI: 10.1038/sj.leu.2403704

[53] Romee R, Foley B, Lenvik T, Wang Y, Zhang B, Ankarlo D. NK cell CD16 surface expression and function is regulated by a disintegrin and metalloprotease-17 (ADAM17). Blood. 2013;**121**(18):3599-3608. DOI: 10.1182/blood-2012-04-425397

[54] Tarazona R, Casado JG, Delarosa O, Torre-Cisneros J, Villanueva JL, Sanchez B, et al. Selective depletion of CD56 dim NK cell subsets and maintenance of CD56 bright NK cells in treatment-naive HIV-1-seropositive individuals. Journal of Clinical Immunology. 2002;**22**(3):176-183

[55] Mavilio D, Lombardo G, Benjamin J, Kim D, Follman D, Marcenaro E, et al. Characterization of CD56–/CD16+ natural killer (NK) cells: A highly dysfunctional NK subset expanded in HIV-infected viremic individuals. Proceedings of the National Academy of Sciences of the United States of America. 2005;**102**(8):2886-2891. DOI: 10.1073/pnas.0409872102

[56] Björkström NK, Ljunggren HG, Sandberg JK. CD56 negative NK cells: Origin, function, and role in chronic viral disease. Trends in Immunology. 2010;**31**(11):401-406. DOI: 10.1016/j.it.2010.08.003

[57] Zulu MZ, Naidoo KK, Mncube Z, Jaggernath M, Goulder PJ, Ndung'u T, et al. Reduced expression of Siglec-7, NKG2A, and CD57 on terminally differentiated $CD56^{-}$ $CD16^{+}$ natural killer cell subset is associated with natural killer cell dysfunction in chronic HIV-1 clade C infection. AIDS Research and Human Retroviruses. 2017;**33**(12):1205-1213. DOI: 10.1089/AID.2017.0095

[58] Sivori S, Carlomagno S, Pesce S, Moretta A, Vitale M, Marcenaro E. TLR/NCR/KIR: Which one to use and when? Frontiers in Immunology. 2014;**5**:105. DOI: 10.3389/fimmu.2014.00105

[59] Mogensen TH. Pathogen recognition and inflammatory signaling in innate immune defenses. Clinical Microbiology Reviews. 2009;**22**(2):240-273. DOI: 10.1128/CMR.00046-08

[60] Qiu F, Maniar A, Diaz MQ, Chapoval AI, Medvedev AE. Activation of cytokine-producing and antitumor activities of natural killer cells and macrophages by engagement of toll-like and NOD-like receptors. Innate Immunity. 2011;**17**(4):375-387. DOI: 10.1177/1753425910372000

[61] Wooldridge JE, Ballas Z, Krieg AM, Weiner GJ. Immunostimulatory

oligodeoxynucleotides containing CpG motifs enhance the efficacy of monoclonal antibody therapy of lymphoma. Blood. 1997;**89**(8):2994-2998

[62] Janeway CA Jr, Medzhitov R. Innate immune recognition. Annual Review of Immunology. 2002;**20**(1):197-216. DOI: 10.1146/annurev.immunol.20.083001.084359

[63] Iwasaki A, Medzhitov R. Toll-like receptor control of the adaptive immune responses. Nature Immunology. 2004;**5**(10):987-995. DOI: 10.1038/ni1112

[64] De Nardo D. Toll-like receptors: Activation, signalling and transcriptional modulation. Cytokine. 2015;**74**(2):181-189. DOI: 10.1016/j.cyto.2015.02.025

[65] Zhang Z, Schluesener HJ. Mammalian toll-like receptors: From endogenous ligands to tissue regeneration. Cellular and Molecular Life Sciences. 2006;**63**(24):2901-2907. DOI: 10.1007/s00018-006-6189-1

[66] Demaria S, Pikarsky E, Karin M, Coussens LM, Chen YC, El-Omar EM, et al. Cancer and inflammation: Promise for biological therapy. J Immunother (Hagerstown, Md.: 1997). 2010;**33**(4):335-351. DOI: 10.1097/CJI.0b013e3181d32e74

[67] Łagiedo M, Sikora J, Kaczmarek M. Damage-associated molecular patterns in the course of lung cancer—A review. Scandinavian Journal of Immunology. 2015;**82**(2):95-101. DOI: 10.1111/sji.12308

[68] Rifkin IR, Leadbetter EA, Busconi L, Viglianti G, Marshak-Rothstein A. Toll-like receptors, endogenous ligands, and systemic autoimmune disease. Immunological Reviews. 2005;**20**(1):27-42. DOI: 10.1111/j.0105-2896.2005.00239.x

[69] Marshak-Rothstein A. Toll-like receptors in systemic autoimmune disease. Nature Reviews Immunology. 2006;**6**(11):823-835. DOI: 10.1038/nri1957

[70] Goh FG, Midwood KS. Intrinsic danger: Activation of toll-like receptors in rheumatoid arthritis. Rheumatology. 2012;**51**(1):7-23. DOI: 10.1093/rheumatology/ker257

[71] Drexler SK, Foxwell BM. The role of toll-like receptors in chronic inflammation. The International Journal of Biochemistry & Cell Biology. 2010;**42**(4):506-518. DOI: 10.1016/j.biocel.2009.10.009

[72] Joosten LA, Abdollahi-Roodsaz S, Dinarello CA, O'Neill L, Netea MG. Toll-like receptors and chronic inflammation in rheumatic diseases: New developments. Nature Reviews Rheumatology. 2016;**12**(6):344-357. DOI: 10.1038/nrrheum.2016.61

[73] Liu Q, Ding JL. The molecular mechanisms of TLR-signaling cooperation in cytokine regulation. Immunology and Cell Biology. 2016;**94**(6):538-542. DOI: 10.1038/icb.2016.18

[74] Roach JC, Glusman G, Rowen L, Kaur A, Purcell MK, Smith KD, et al. The evolution of vertebrate Toll-like receptors. Proceedings of the National Academy of Sciences of the United States of America. 2005;**102**(27):9577-9582. DOI: 10.1073/pnas.0502272102

[75] Uematsu S, Akira S. Toll-like receptors and type I interferons. The Journal of Biological Chemistry. 2007;**282**(21):15319-15323. DOI: 10.1074/jbc.R700009200

[76] Gaudreault E, Fiola S, Olivier M, Gosselin J. Epstein-Barr virus induces MCP-1 secretion by human monocytes via TLR2. Journal of Virology. 2007;**81**(15):8016-8024. DOI: 10.1128/JVI.00403-07

[77] Hertz CJ, Wu Q, Porter EM, Zhang YJ, Weismüller KH, Godowski PJ, et al. Activation of toll-like receptor 2 on human tracheobronchial epithelial cells induces the antimicrobial peptide human β defensin-2. Journal of Immunology. 2003;**171**(12):6820-6826. DOI: 10.4049/jimmunol.171.12.6820

[78] Guillot L, Le Goffic R, Bloch S, Escriou N, Akira S, Chignard M, et al. Involvement of toll-like receptor 3 in the immune response of lung epithelial cells to double-stranded RNA and influenza a virus. The Journal of Biological Chemistry. 2005;**280**(7):5571-5580. DOI: 10.1074/jbc.M410592200

[79] Vivier E, Biron CA. A pathogen receptor on natural killer cells. Science. 2002;**296**(5571):1248-1249. DOI: 10.1126/science.1072447

[80] Kadowaki N, Antonenko S, Ho S, Rissoan MC, Soumelis V, Porcelli SA, et al. Distinct cytokine profiles of neonatal natural killer T cells after expansion with subsets of dendritic cells. The Journal of Experimental Medicine. 2001;**193**(10):1221-1226

[81] Uehori J, Matsumoto M, Tsuji S, Akazawa T, Takeuchi O, Akira S, et al. Simultaneous blocking of human Toll-like receptors 2 and 4 suppresses myeloid dendritic cell activation induced by Mycobacterium bovis bacillus Calmette-Guerin peptidoglycan. Infection and Immunity. 2003;**71**(8):4238-4249. DOI: 10.1128/IAI.71.8.4238-4249.2003

[82] Cella M, Salio M, Sakakibara Y, Langen H, Julkunen I, Lanzavecchia A. Maturation, activation, and protection of dendritic cells induced by double-stranded RNA. The Journal of Experimental Medicine. 1999;**189**(5):821-829

[83] Adib-Conquy M, Scott-Algara D, Cavaillon JM, Souza-Fonseca-Guimaraes F. TLR-mediated activation of NK cells and their role in bacterial/viral immune responses in mammals. Immunology and Cell Biology. 2014;**92**(3):256-262. DOI: 10.1038/icb.2013.99

[84] Hayden MS, Ghosh S. NF-κB in immunobiology. Cell Research. 2011;**21**(2):223-244. DOI: 10.1038/cr.2011.13

[85] Hornung V, Rothenfusser S, Britsch S, Krug A, Jahrshdörfer B, Giese T, et al. Quantitative expression of toll-like receptor 1-10 mRNA in cellular subsets of human peripheral blood mononuclear cells and sensitivity to CpG oligodeoxynucleotides. Journal of Immunology. 2002;**168**(9):4531-4537. DOI: 10.4049/jimmunol.168.9.4531

[86] He S, Chu J, Wu LC, Mao H, Peng Y, Alvarez-Breckenridge CA, et al. MicroRNAs activate natural killer cells through toll-like receptor signaling. Blood. 2013;**121**(23):4663-4671. DOI: 10.1182/blood-2012-07-441360

[87] Saikh KU, Lee JS, Kissner TL, Dyas B, Ulrich RG. Toll-like receptor and cytokine expression patterns of $CD56^+$ T cells are similar to natural killer cells in response to infection with Venezuelan equine encephalitis virus replicons. The Journal of Infectious Diseases. 2003;**188**(10):1562-1570. DOI: 10.1086/379196

[88] Chalifour A, Jeannin P, Gauchat JF, Blaecke A, Malissard M, N'Guyen T, et al. Direct bacterial protein PAMP recognition by human NK cells involves TLRs and triggers α-defensin production. Blood. 2004;**104**(6):1778-1783. DOI: 10.1182/blood-2003-08-2820

[89] Gorski KS, Waller EL, Bjornton-Severson J, Hanten JA, Riter CL, Kieper WC, et al. Distinct indirect pathways govern human NK-cell activation by TLR-7 and TLR-8 agonists. International Immunology. 2006;**18**(7):1115-1126. DOI: 10.1093/intimm/dxl046

[90] Alter G, Suscovich TJ, Teigen N, Meier A, Streeck H, Brander C, et al. Single-stranded RNA derived from HIV-1 serves as a potent activator of NK cells. Journal of Immunology. 2007;**178**(12):7658-7666. DOI: 10.4049/jimmunol.178.12.7658

[91] Lauzon NM, Mian F, MacKenzie R, Ashkar AA. The direct effects of toll-like receptor ligands on human NK cell cytokine production and cytotoxicity. Cellular Immunology. 2006;**241**(2):102-112. DOI: 10.1016/j.cellimm.2006.08.004

[92] Siednienko J, Miggin SM. Expression analysis of the toll-like receptors in human peripheral blood mononuclear cells. Methods in Molecular Biology. 2009;**517**:3-14. DOI: 10.1007/978-1-59745-541-1_1

[93] Girart MV, Fuertes MB, Domaica CI, Rossi LE, Zwirner NW. Engagement of TLR3, TLR7, and NKG2D regulate IFN-γ secretion but not NKG2D-mediated cytotoxicity by human NK cells stimulated with suboptimal doses of IL-12. Journal of Immunology. 2007;**179**(6):3472-3479. DOI: 10.4049/jimmunol.179.6.3472

[94] Hart OM, Athie-Morales V, O'Connor GM, Gardiner CM. TLR7/8-mediated activation of human NK cells results in accessory cell-dependent IFN-γ production. Journal of Immunology. 2005;**175**(3):1636-1642. DOI: 10.4049/jimmunol.175.3.1636

[95] Souza-Fonseca-Guimaraes F, Parlato M, Philippart F, Misset B, Cavaillon JM, Adib-Conquy M. Toll-like receptors expression and interferon-γ production by NK cells in human sepsis. Critical Care. 2012;**16**(5):R206. DOI: 10.1186/cc11838

[96] Martinez J, Huang X, Yang Y. Direct TLR2 signaling is critical for NK cell activation and function in response to vaccinia viral infection. PLoS Pathogens. 2010;**6**(3):e1000811. DOI: 10.1371/journal.ppat.1000811

[97] Cañeda-Guzmán IC, Salaiza-Suazo N, Fernández-Figueroa EA, Carrada-Figueroa G, Aguirre-García M, Becker I. NK cell activity differs between patients with localized and diffuse cutaneous leishmaniasis infected with leishmania mexicana: A comparative study of TLRs and cytokines. PLoS ONE. 2014;**9**(11):e112410. DOI: 10.1371/journal.pone.0112410. eCollection 2014

[98] Lindgren Å, Pavlovic V, Flach CF, Sjöling Å, Lundin S. Interferon-gamma secretion is induced in IL-12 stimulated human NK cells by recognition of *Helicobacter pylori* or TLR2 ligands. Innate Immunity. 2011;**17**(2):191-203. DOI: 10.1177/1753425909357970

[99] Kanevskiy L, Telford WG, Sapozhnikov AM, Kovalenko EI. Lipopolysaccharide induces IFN-γ production in human NK cells. Frontiers in Immunology. 2013;**4**:1-10. DOI: 10.3389/fimmu.2013.00011

[100] Roda JM, Parihar R, Carson WE. CpG-containing oligodeoxynucleotides act through TLR9 to enhance the NK cell cytokine response to antibody-coated tumor cells. Journal of Immunology. 2005;**175**(3):1619-1627. DOI: 10.4049/jimmunol.175.3.1619

[101] Merlo A, Calcaterra C, Mènard S, Balsari A. Cross-talk between Toll-like receptors 5 and 9 on activation of human immune responses. Journal of Leukocyte Biology. 2007;**82**(3):509-518. DOI: 10.1189/jlb.0207100

[102] Tsujimoto H, Uchida T, Efron PA, Scumpia PO, Verma A, Matsumoto T, et al. Flagellin enhances NK cell proliferation and activation directly and through dendritic cell-NK cell interactions. Journal of Leukocyte Biology. 2005;**78**(4):888-897. DOI: 10.1189/jlb.0105051

[103] Vacchelli E, Prada N, Kepp O, Galluzzi L. Current trends of

anticancer immunochemotherapy. OncoImmunology. 2013;**2**(6):e25396. DOI: 10.4161/onci.25396

[104] Lu H. TLR agonists for cancer immunotherapy: Tipping the balance between the immune stimulatory and inhibitory effects. Frontiers in Immunology. 2014;**5**:83. DOI: 10.3389/fimmu.2014.00083

[105] Li K, Qu S, Chen X, Wu Q, Shi M. Promising targets for cancer immunotherapy: TLRs, RLRs, and STING-mediated innate immune pathways. International Journal of Molecular Sciences. 2017;**18**(2):E404. DOI: 10.3390/ijms18020404

Chapter 3

TLR-Mediated Host Immune Response to Parasitic Infectious Diseases

M. Magdalena Aguirre-García, Araceli Rojas-Bernabé, A. Pamela Gómez-García and Alma R. Escalona-Montaño

Abstract

Toll-like receptors (TLRs) are important for the host immune response to a variety of pathogens, including bacteria, viruses, fungi, and parasites. These receptors become activated upon recognizing pathogen-associated molecular patterns (PAMPs) and thus initiate the innate immune response to the corresponding pathogen. A key aspect of TLRs is their activation of signaling that leads to cytokine production and an inflammatory response. Additionally, TLRs act as the bridge between innate and acquired immunity, enhancing phagocytosis and the process of killing parasites. We herein focus on how parasites (protozoans and helminths) and their derived products have the capability of stimulating or evading the host response by triggering or inhibiting TLR activation. Parasites often develop successful survival strategies that imply interference with the host immune response. Accordingly, many of these organisms have molecules that modulate inflammation and other aspects of host immunity. Taking advantage of such mechanisms, there are some anti-inflammatory therapies based on human infection with helminths. Helminths and protozoans influence the activity of various TLRs, especially TLR2, TLR4, and TLR9. A better understanding of the role of TLRs and their parasite-derived ligands should certainly provide new therapeutic tools for combatting various parasitic and inflammatory diseases.

Keywords: TLRs, protozoans, helminths, immune response, disease

1. Introduction

Toll-like receptors (TLRs) have the important function of recognizing a variety of pathogens, including bacteria, viruses, fungi, and parasites. They recognize pathogen-associated molecular patterns (PAMPs) and consequently initiate the innate immune response. The well-known TLRs that are responsible for binding to PAMPs act as a bridge between innate and acquired immunity. Accordingly, they are not only involved in the production of cytokines and chemokines but also enhance phagocytosis and the process of killing parasites.

Parasites are organisms that live on or inside an organism and benefit by deriving nutrients at the expense of the host. We herein review how protozoan and helminth parasites trigger differential activation of TLRs in order to regulate host immune cells. In some cases, the activation of these TLR receptors by PAMPs contributes to an effective control of the infection, while in other cases there is a negative regulation resulting in the exacerbation of the infection, as shown in distinct in vivo animal models. In patients with diverse clinical manifestations of parasitic infections, TLRs and cytokines play a key role in the host response to the disease. A better understanding of the role of TLRs and their ligands should certainly provide new therapeutic tools and perhaps allow for the development of vaccines to control parasitic infections.

2. Protozoan parasites

Among the many protozoan parasites, those presently examined are *Leishmania* spp., *Trypanosoma* spp., *Naegleria fowleri, Plasmodium* spp., *Toxoplasma gondii*, *Giardia lamblia*, *Entamoeba histolytica*, *Trichomonas vaginalis*, *Blastocystis* spp., and *Acanthamoeba* spp.

2.1 *Leishmania* spp.

Leishmaniasis, an infectious disease endemic in 88 countries representing 5 continents, is caused by parasitic protozoa in the genus *Leishmania* [1]. This parasite gives rise to a variety of disorders, ranging from cutaneous lesions to visceral disease [2]. The latter can be generated by *Leishmania donovani*, *L. infantum*, and *L. chagasi* [3].

Regarding cutaneous lesions, the most pathogenic agents of American cutaneous leishmaniasis in Brazil are *L. (Viannia) braziliensis* and *L.* (L.) *amazonensis*, capable of inducing localized cutaneous leishmaniasis, borderline disseminated cutaneous leishmaniasis, anergic diffuse cutaneous leishmaniasis, and mucosal leishmaniasis [4]. In Mexico, *L. mexicana* evokes a wide spectrum of cutaneous diseases. For instance, localized cutaneous leishmaniasis is characterized by ulcers at the site of parasite inoculation, while parasites in diffuse cutaneous leishmaniasis spread throughout the skin and form disfiguring nodules [5]. In Iran, cutaneous leishmaniasis is endemic in 18 of 31 provinces, and approximately one-fifth of the cases belong to anthroponotic cutaneous leishmaniasis, stemming from *L. tropica* [6]. *L. panamensis*, a member of the *Viannia* subgenus of *Leishmania*, is known to provoke mucosal leishmaniasis. It produces destructive lesions of the nasal, oral, and hypopharyngeal mucosa [1].

2.2 *Trypanosoma* spp.

The causal agent of Chagas disease, *Trypanosoma cruzi*, was first described by the Brazilian physician Carlos Chagas in 1909 [7]. This pathology is endemic in Central and South America, and evidence exists of some cases in the United States, Europe, and Japan due to travel and migration. The parasite is an intracellular protozoan of the Trypanosomatidae family, transmitted to humans by blood-feeding reduviid bugs.

There are two phases of Chagas disease. The acute phase is generally asymptomatic, although some patients present symptoms such as fever, nausea, vomiting, anorexia, and diarrhea [8]. In the chronic phase, infected individuals can remain asymptomatic for decades, although around 30% eventually develop cardiac or gastrointestinal complications characteristic of the disease.

2.3 Other protozoans

Naegleria fowleri is a protozoan that invades the central nervous system and provokes primary amoebic meningoencephalitis. During the process of infection, it induces an important inflammatory response [9].

Acanthamoeba spp. are free-living amoebae found in lakes, rivers, swimming pools, thermal baths, and tap water [10]. They infect humans and animals as opportunistic pathogens in immunocompromised hosts [11]. These parasites are able to generate severe diseases, including amebic *Acanthamoeba keratitis*, a painful sight-threatening infection of the cornea, and granulomatous amebic encephalitis, a fatal disease of the central nervous system.

Giardia lamblia, the causal agent of giardiasis, colonizes the lumen of the upper small intestine. The parasite adheres to the surface of enterocytes without traversing the enterocyte barrier [12].

P. falciparum, the protozoan parasite responsible for malaria, is transmitted by the bite of mosquitoes. It results in a wide spectrum of clinical manifestations during the vector-parasite-host interaction. The first asexual reproduction process occurs in the human liver [13].

Trichomonas vaginalis is a flagellated protozoan parasite that infects the human genitourinary tract. It is the causative organism of trichomoniasis, one of the most prevalent sexually transmitted diseases in the world [14].

Entamoeba histolytica, the etiologic agent of amebiasis, is a pathogenic enteric protozoan. The manifestations of the disease range from mild diarrhea to severe dysentery, with liver abscesses forming in rare cases [15].

Blastocystis, an enteric parasite, colonizes the colonic epithelia of human and animal hosts [16]. Infection with this parasite gives rise to diarrhea, abdominal pain, flatulence, vomiting, and bloating [17].

3. Helminth parasites

The word helminth is derived from the Greek "helmins," which means parasite worm. Helminth is an umbrella term that includes many species of worms from different genera, having parasitism in common. They are quite frequently found in the population. The immune response depends largely on the type of parasite and its interaction with the host.

These large extracellular organisms have a complicated life cycle. They frequently migrate through blood vessels and tissues until reaching the definitive organ, such as the intestine, lungs, liver, or lymphatic organs. Some invade and colonize various cell types.

Because helminths have developed strategies of evasion of the host immune response, they are often able to survive. Hence, they can weaken the immune response and survive for years in the infected host, establishing chronic infection [18]. When the host immune response is adequately activated, on the other hand, infections are eliminated quickly.

Fasciola hepatica is a trematode worm that mainly affects livestock (e.g., cows, sheep, and goats) and humans. It is transmitted through the ingestion of aquatic plants contaminated with metacercaria. *Ascaris lumbricoides*, a parasitic intestinal worm, is transmitted by ingesting water contaminated with embryonic eggs. *Trichuris trichiura* and *T. suis*, other species of intestinal worms, remain in the large intestine of mammals in the form of adult larvae. The infection occurs after the ingestion of the eggs, which hatch in the small intestine and release the infective larvae [19]. *Schistosoma mansoni* are larvae of worms that live in waters and ponds contaminated by feces. They are able to penetrate the skin of people who bath or swim in contaminated pools. In the adult stage, these parasites produce eggs that are excreted through the stool [20]. *Strongyloides stercoralis* is an infection generated by the larvae. These are acquired through the skin and later lodge themselves in the intestine. *Trichinella spiralis* is a parasitic nematode, which infects the muscle tissue of practically all mammals. *Toxocara canis*, round worms found in dogs, can infect humans if ingested in the form of the infective eggs of *Toxocara*. The larvae migrate to the intestine, liver, or lungs.

4. Molecules and protozoan and helminth parasites activate through TLRs

4.1 *Leishmania* spp. protozoans

L. major prompts the activation of the IL-1β promoter and mRNA expression in macrophages through the MyD88-pathway [21]. Additionally, *L. major* LPG upregulates the expression of TLR2 in NK cells and the production of cytokines, such as IFN-γ, TNF-α, and nuclear factor NFκB [22].

Amastigotes from *L.* (*V.*) *braziliensis* reduce the expression of TLR4 on the membrane of macrophages. This receptor modulates the production of TNF-α and IL-10 [23]. *L. infantum* promastigotes stimulate the production of IFN-α/β in plasmacytoid dendritic cells and the release of IL-12 by myeloid dendritic cells. Both these cytokines are dependent on TLR9 [24]. The expression of TLR9 and the production of TNF-α and IL-12 were observed in macrophages exposed to DNA from *L. mexicana* promastigotes [25]. Similarly, DNA from *L. major* was identified as the specific ligand that triggered TLR9-dependent activation of dendritic cells [26]. On the other hand, antigens of *L. donovani* stimulated the production of TNF-α, IL-12, and IFN-γ and increased TLR2 gene expression on RAW264.7 macrophages [27].

L. panamensis infection upregulates the expression of TLR1, TLR2, TLR3, and TLR4 and the production of TNF-α in human primary macrophages [28]. In peripheral blood mononuclear cells, *L. mexicana* LPG fostered the production of TNF-α, IL-1β, IL-12 p40, IL-12 p70, and IL-10 and the expression of TLR2 and TLR4. The triggering of the latter TLRs led to phosphorylation of the extracellular regulated kinase (ERK) and p38 mitogen-activated protein kinase (MAPK) [29].

4.2 Other protozoans

Glycosylphosphatidylinositol (GPI), a molecule derived from trypomastigotes, was found to be a potent activator of TLR2 from human and mouse origin [30]. Another family of GPIs, glycoinositolphospholipid (GIPL), has been purified from epimastigotes and triggers NF-κB activation via TLR4 [31].

Whether the recognition of different GPI anchors is mediated by TLR2 or TLR4 depends on their variable lipid moiety composition. There is one report of RA (high virulence) and K 98 (low virulence) strains of *T. cruzi* used to obtain total lipid extracts. The authors demonstrated that the total lipids from both strains promote the formation of lipid bodies and the release of pro-inflammatory molecules, including cyclooxygenase-2, TNF-α, and nitric oxide (NO) in macrophages as well as HEK cells, all through a TLR2/6-dependent pathway [32].

P. falciparum GPI contributes to the pathology of malaria by inducing the release of cytokines. In vitro studies have shown that this molecule is recognized by TLR2 and to a lesser degree by TLR4 [33]. Hemozoin stimulates the production of IFN-γ and inflammatory cytokines by triggering TLR9 in dendritic cells [34].

T. vaginalis parasites enhance TLR2 gene expression and activate NOD-like receptor type 3 (NLRP3) inflammasome in mouse macrophages and THP-1 human macrophages, respectively [35]. They also regulate the production of pro-inflammatory cytokines by the activation of MAPK and NFκB p65 and prompt pyroptotic cell death through the release of IL-1β [36]. TLR4 upregulation has been reported in a prostate stromal cell line exposed to *T. vaginalis* [37].

Profiling like molecules and heat shock protein 70 from *T. gondii* activates dendritic cells through TLR4 and TLR11 [38]. The lipopeptidophosphoglycan (LPPG) and DNA from *E. histolytica* are recognized by TLR2, TLR4, and TLR9, triggering the release of IL-10, IL-12p40, TNF-α cytokines, and IL-8 from human monocytes [39, 40]. Unmethylated CpG oligodeoxynucleotides (CpG ODN) from the same parasite generate MMP-9 expression via the TLR9-dependent activation of ERK and p38 MAPK followed by the activation of NFκB [41]. *G. lamblia* trophozoites trigger TLR2, resulting in the activation of ERK and MAPK and the production of pro-inflammatory cytokines in peritoneal macrophages of wild-type mice [42]. Live *Blastocystis* spp. parasites and whole cell lysate alone cannot trigger TLRs in THP-1 human monocytes. ST4WR1 parasites inhibit LPS-mediated activation of NF-κB in these same cells [43]. *N. fowleri* elicits the expression and production of pro-inflammatory cytokines and β-defensing-2, mainly through the canonical TLR4 pathway in a time-dependent manner [44].

4.3 Helminths

The host immune response mounted against helminth infections is activated through TLRs, which are triggered by the glycoproteins, secretion/excretion products, and various other molecules of the parasites [45].

Cathepsin cysteine protease (FheCL1) of *F. hepatica* inhibits the secretion of various inflammatory mediators, such as TNF-α, IL-12, and NO [46]. This is carried out by the null activation of macrophages through the degradation of TLR3 in the endosome in a TRIF-dependent pathway independent of MyD88 [47]. Glucans promote the production of high levels of IL-10 and IL-4, thus favoring a Th2 response. These glycoconjugates modulate the function and maturation of dendritic cells. In the process of phase changes, *Fasciola* employs an immunosuppressive mechanism that favors a Th2 response [48], in part by releasing various excretory/secretory substances that inhibit the maturation of dendritic cells and trigger TLRs via the MyD88-dependent signaling pathway.

The excretory/secretory molecules of *Taenia crassiceps* initiate the phosphorylation of cRAF by means of MGL (lectins) and TLR2, as well as decreasing the maturation of dendritic cells and the production of IL-12 and TNF-α. These molecules also modulate the signaling pathways of NFκB p65 and p38 MAPK activated by LPS through TLR4 [49, 50] and regulate the type C receptor of lectin [51]. The carbohydrates of the parasite induce the production of IL-6 through TLRs [52].

On the other hand, the lysophosphatidylserine and lipopolysaccharide of *A. lumbricoides* generate signaling through TLR2, modulate a Th2 response (leading to the secretion of IL-10), and promote phosphorylation of ERK 1/2. This helminth binds to hyaluronic acid (the main constituent of connective tissue), activates dendritic cells by TLR4 (in adult parasites, or in larvae depending on the larval concentration), and provokes inflammatory processes. Through its antioxidant properties, it can eliminate free radicals and act as a barrier to tissue degradation.

Trichuris trichiura inhibits the function of TLR4 by suppressing LPS-induced production and the secretion of TNF-α [53]. Similarly, cathepsin B1 secreted by *S. mansoni* inhibits the recognition of TLR3 and TLR4 by inactivating the MyD88-independent pathway and modulating Th2 responses. Phospholipids and glycophosphatidylserine trigger TLR2 in dendritic cells, eliciting a Th2 response and the stimulation of Tregs to secrete IL-10.

The soluble antigen of *S. mansoni* suppresses the production of cytokines IL-1β, IL-6, IL-12, and TNF-α as well as increases the secretion of TGF-β and the production of IL-10 in dendritic cells. Activated B cells augment the production of IL-10.

Both processes are dependent on the triggering of TLR2 and activation of MyD88, the latter of which promotes ERK1/2 phosphorylation [20]. dsRNA from *S. mansoni* binds to TLR3 in dendritic cells, thus positively regulating the expression of IFN type 1 genes [54–56].

Although the TLR4 receptor is not essential for killing the larva of *S. stercoralis* during the innate immune response, it is indeed crucial for killing the parasites during the adaptive immune response [57].

Phosphorylcholine is a glycoprotein released by *T. spiralis* that binds to TLR4 [58]. Several TLRs decrease the susceptibility of effector T cells to Treg-mediated suppression [59]. During the first week of infection, the gene expression of TLR1,

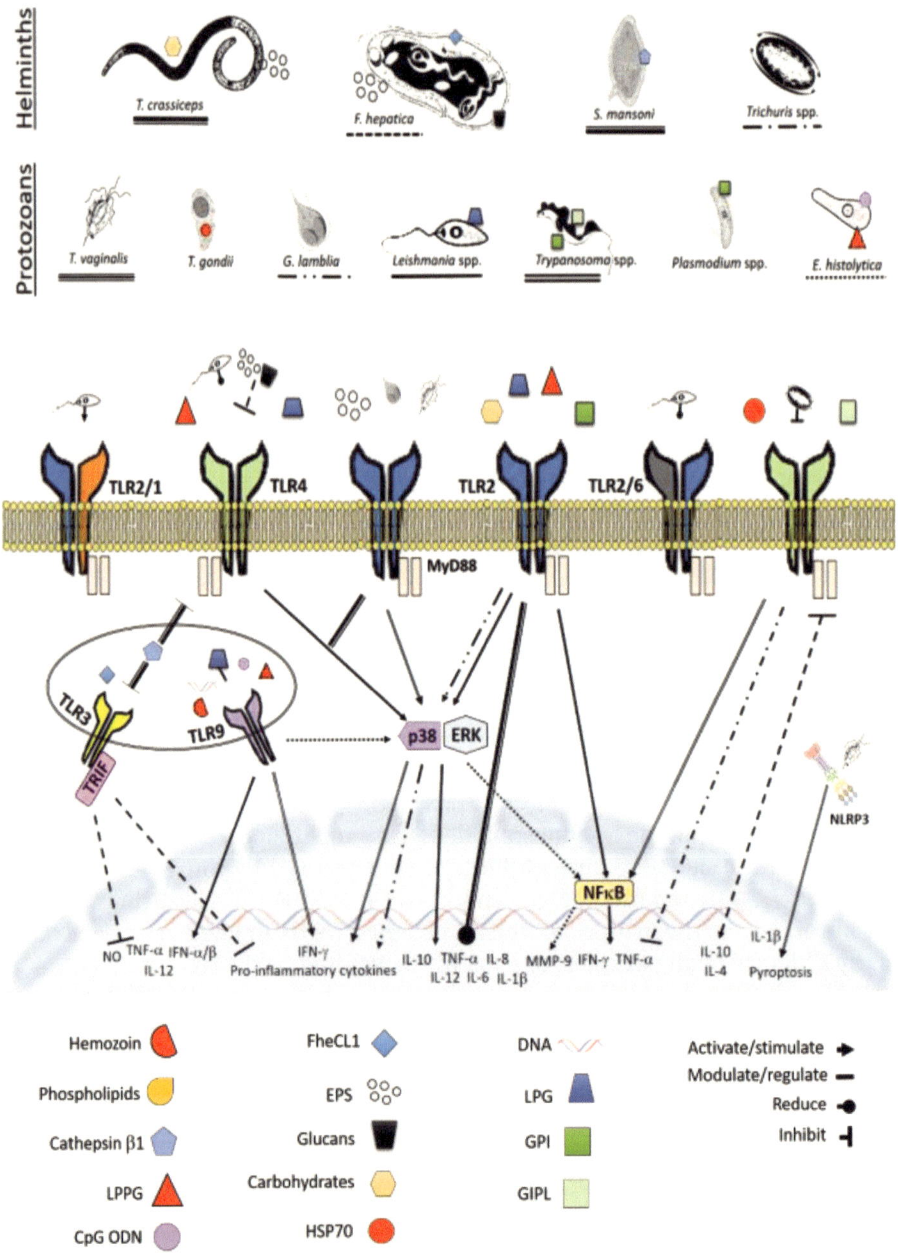

Figure 1.
TLR activation by PAMPs of parasites.

TLR2, TLR4, TLR5, and TLR9 increases significantly in the intestinal and muscular phases [60]. Infection also activates dendritic cells through TLR2 and TLR4 [61]. Since the TLR4/MyD88/NF-κB signaling pathway affects the secretion of inflammatory cytokines in macrophages, it may participate in immune suppression. In this sense, *T. spiralis* infection modulates the expression of TLR2 and TLR4 during different stages. Cytokine levels are regulated via TLR4-mediated signaling pathway, suggesting that TLR4 modulates host immunosuppression during infection [62].

Infection with *Toxocara* prompts cells to produce cytokines such as IL-4, IL-5, and IL-13, which generate increased IgE levels [63]. *Toxocara canis* inhibits the response of TLR2 to the activation of dendritic cells and competes with the host for lectins, thus blocking host immunity [64]. The activation of TLRs pathways by PAMPs of protozoans and helminths is summarized in **Figure 1**.

5. TLRs contribute to the effective control of parasite infection

5.1 *Leishmania* spp.

Infection by *L. major* in TLR4-deficient mice proliferates and correlates with a higher activity of arginase. In contrast, the same infection in TLR4-competent mice leads to low parasite replication that correlates with higher levels of inducible nitric oxide synthase. Hence, TLR4 competence could resolve cutaneous lesions and control parasite growth [65, 66].

TLR2 is unique among TLRs because it forms heterodimers with TLR1 or TLR6 and modulates downstream signaling pathways. In peritoneal macrophages from BALB/c mice infected with *L. major* promastigotes, the expression of TLR1 and TLR2 increases but not that of TLR6. TLR2-TLR2 association increases but TLR2-TLR6 association diminishes. The disparity in the formation of heterodimers between TLR2 and TLR1 or TLR2 and TLR6 brings about a TLR2 functional duality [67]. Additionally, LPG from *L. major* participates in the modulation of TLR9 expression and function.

The question arises as to whether the functional duality of TLR2 contributes to such modulation. The co-administration of CpG and the anti-TLR2 antibody reduces infection in susceptible BALB/c mice, establishing a relationship between LPG, TLR2, and TLR9 during an *L. major* infection [27]. In C57BL/6 mice deficient in either TLR2, 4, or 9, only the TLR9-deficient animals are more susceptible to infection with *L. major*. The deficiency of TLR9 inhibits the development of the curative Th1 response [26]. This effect has also been observed in TLR9-deficient mice infected with *L. infantum*. Therefore, TLR9 appears to play a key role in infections by *L. major* or *L. infantum* [2]. On the other hand, TLR11 and TLR12 play an important role in an *L. major* infection, evidenced by the fact that the silencing of these receptors reduces the parasite burden, enhances the level of IFN-γ, and diminishes the production of IL-4 [68].

Bone marrow-derived dendritic cells of $MyD88^{-/-}$ and $TLR2^{-/-}$ mice were infected with *L. braziliensis.* Compared to control mice, the infected dendritic cells from $MyD88^{-/-}$ mice showed low levels of activation and decreased production of cytokine IL-12p40, leading to greater infection by *L. braziliensis* and a limited expansion of $CD4^{+}$ T cells that produce IFN-γ and IL-17 during infection. In contrast, $TLR2^{-/-}$ mice were more resistant to infection than control mice, which was associated with increased production of IFN-γ. Thus, MyD88 seems to be essential for the recognition of *L. braziliensis*, and TLR2 apparently has a regulatory role in modulating the immune response to the same parasite [69]. In peripheral blood mononuclear cells from cutaneous leishmaniasis patients infected with *L. braziliensis*, an evaluation was made of the expression of co-stimulatory molecules CD80,

CD86, and TLR9. Monocytes decreased the expression of these co-stimulatory molecules, but their expression was higher when the cells were exposed to soluble *Leishmania* antigen [70].

In the spleen of mice infected with *L. chagasi*, possible correlations were analyzed between TLR2 and TLR4 mRNA expression and the production of cytokines and nitric oxide (NO). Whereas the mRNA expression of TLR2, TLR4, IL-17, TNF-α, and TGF-β increases in the early stage of infection, it decreases during the late stage, which correlates with parasite load. The mRNA expression of IFN-γ and IL-12 declines at the peak of infection [3].

On the other hand, mice with the A20 protein silenced (a protein that inhibits NFκB activation by modulating the function of IKK, TRAF6, and NEMO) were infected with *L. donovan*, finding enhancement of the activation of NFκB and the host-protective pro-inflammatory response (compared to normal mice), which correlates with effective parasite clearance [71].

5.2 Other protozoans

Single-nucleotide polymorphisms have a role in innate immune responses and in susceptibility/resistance to infection. However, the genetic association of TLR4 and TNF-α polymorphisms is a determinant factor for the development of disease in Chagas disease. It has been reported the greatest parasite load exists in individuals with the C/C-Asp/Asp-Thr/Thr combination of homozygous genotypes at the TNF-α promoter 1031 and on the TLR4 polypeptide. The former homozygous genotype is represented by C/C. The latter homozygous genotypes encode Asp/Asp and Thr/Thr at codons 299 and 399, respectively [72]. Two TLR1 polymorphisms, rs4833095 (Asn248Ser) and rs5743618 (Ser602Ile), were assessed among 302 primiparous Ghanaian women for their association with *P. falciparum* infection and placental malaria and susceptibility to placental malaria [73]. It was found that TLR4-Asp299Gly and the TLR4-Thr399Ile variants confer greater risk of severe malaria in Ghanaian children [74]. The single-nucleotide polymorphism of TLR6 S249P may be a risk factor for the development of malaria [75]. Regarding TLR5, the single-nucleotide polymorphism of the TIR domain (TIRAP) S180 L provides protection against malaria, while the single-nucleotide polymorphism of R392 stop codon increases susceptibility to the same disease [76]. An rs4986790 polymorphism of the TLR4 gene can modulate the susceptibility toward a *P. vivax* infection. The AA genotype proved to be protective against the development of this parasite in a local population of Pakistan [77]. Additionally, TLR4 A299G, TLR6 S249, and TLR 9-1486C/T influence the levels of circulating cytokines IL-6, IFN-γ, IL-12, IL-10, and IL-4 during a *P. vivax* infection [78].

Mice deficient in the adaptor protein MyD88 and IL-18 were more susceptible to *P. yoelii* infection and increased parasitemia during the early phase of the infection. Greater lethality was observed in the MyD88-deficient animals. In mice deficient in IL-1R, parasitemia showed a slight increment during infection with *P. yoelii* [79]. During infection with the same parasite, there was a high level of parasite burden in $TLR2^{-/-}$ mice, which was closely associated with a reduction in pro-inflammatory cytokines in the liver [80].

On the other hand, the TLR9 variant −1237C/C correlates with acute parasitemia during a *P. vivax* infection [81]. An endoplasmic reticulum resident protein, UNC93B1, is critical for host resistance to *T. cruzi* and *T. gondii* [80, 82]. Its function is the translocation of nucleotide-sensing TLRs from the endoplasmic reticulum to endolysosomes. Interestingly, $TLR2^{-/-}$ and AKT-blocked mice that are infected with *Giardia* display a decreased parasite burden, an increased weight gain rate, and a shorter parasite persistence compared to normal mice infected with the same protozoan [42].

5.3 Helminths

The immune responses against helminth infections do not efficiently protect the host. The immune system is unable to eliminate chronic infection, and the immune memory fails to protect against reinfection, even after a cure mediated by pharmacological treatment [83].

For an *S. mansoni* or *A. lumbricoides* infection, dendritic cells are fundamental in the protection and activation of host defensive responses. When TLR receptors on the surface of dendritic cells bind to the ligands of the *S. mansoni*, *A. lumbricoides*, and *T. trichiura* parasites, there is an increased expression of co-stimulatory molecules (CD40, CD80, and CD86) and synthesis of pro-inflammatory mediators, such as IL-12 and TNF-α, that evoke a Th1 lymphocyte response [84, 85]. For intestinal parasites, the mRNA expression of TLR1, TLR2, TLR3, TLR4, and TLR9 is regulated in the early stages of infection. During the adult stage of an infection with *T. spiralis*, TLR1 and TLR4 activate the signaling pathway dependent on MyD88. Once newborn larvae appear, however, all expression of TLRs is inhibited, except for TLR2. The expression of TLR2/4 in the small intestine and muscle tissue during infection could be closely associated with immune responses mediated by Treg cells and greater expression of cytokines such as IL-10 and TGF-β [86, 87]. During an infection by *S. mansoni*, TLR2 and TLR4 limit the activation of the intestinal immune response. Consequently, the deficiency of these TLRs promotes the expulsion of adult worms. In the course of the first weeks of infection of animals with such deficiency, the expression of several TLRs increases, an active state that begins to decrease as of the fourth week. This favors an early activation of the immune response in the host and the prompt expulsion of the parasites [88].

6. Negative regulation by TLRs

6.1 *Leishmania donovani*, *Giardia lamblia*, and *Schistosoma mansoni*

L. donovani infection results in a suppression of TLR2- and TLR4-stimulated production of IL-12p40 and an increase in IL-10 production. Additionally, the parasites modulated the MAPK pathway by suppressing TLR2-dependent MAPK and ERK phosphorylation [89].

An in vitro infection with *G. lamblia* leads to a decreased production of pro-inflammatory cytokines by activating the AKT signaling pathway via TLR2.

Helminths are regulated by the expression of soluble antigens, which in parasites such as *S. mansoni* and *F. hepatica* exert an inhibitory effect on the maturation of dendritic cells induced by TLR ligands. The dendritic cells activated by soluble antigens produce a smaller amount of IL-12 than those activated only with the ligand for TLR4 [90–92]. The soluble antigen extracts of *S. mansoni* inhibit the ability of CpG, poly I:C, hyaluronic acid, and LPS to stimulate the production of IL-12 or increase the surface expression of CD80, CD86, and MHC class II in dendritic cells. Therefore, these extracts decrease the production of IL-12, IL-6, and TNF-α and the expression of co-stimulatory CD80/86 molecules, which in turn suppresses the Th1 response and favors the Th2 response elicited by LPS [93, 94]. The negative regulation of TLRs can reduce the production of pro-inflammatory cytokines, which could protect the host from autoimmune pathogenesis. The mechanism of this action is the regulation of the Th1/2 cell balance and the modulation of TLR4 expression [92, 95, 96].

In order to adapt to the immune system, some helminth parasites activate and/or downregulate TLRs, as well as interfere with the expression of several genes related to the transduction pathway [86]. Several studies suggest that continuous exposure to helminth antigens may negatively regulate the response of cells to PAMPs derived from these parasites, implying a weakened immune response in individuals infected with helminths.

6.2 Highly virulent parasites downregulate TLR expression

According to an in vitro assay, *T. cruzi* strains with low virulence cause relatively high expression of TLR4 and high levels of pro-inflammatory cytokines such as IL-12 and TNF-α. Contrarily, virulent *T. cruzi* strains maintain a low expression of TLR4 and reduced production of TNF-α [97]. Likewise, at 2 days post-infection with *Acanthamoeba* strains Ac55 and Ac43, the mRNA expression of TLR2 and TLR4 was elevated in the brain of mice (versus control animals) [98]. Compared to uninfected mice, moreover, *Acanthamoeba* sp. generated a high expression of the mRNA and high levels of *TLR2* in the brain of animals at 2, 4, 8, 16, and 30 days post-infection [99].

7. TLRs in the outcome of disease

7.1 Leishmaniasis

Splenic biopsies and peripheral blood samples (to obtain mononuclear cell isolates) were taken from patients with visceral leishmaniasis in India, detecting the mRNA expression of TLR2 and TLR4 but not TLR9. The mRNA expression of IL-10 and IFN-γ was greater in pre-treatment versus posttreatment splenic biopsies. Additionally, the levels of IFN-γ and IL-10 mRNA were higher in peripheral blood mononuclear cells for pre-treatment versus posttreatment patients and healthy controls [100].

In Brazil, the peripheral blood mononuclear cells and lymphocytes ($CD14^+$ and $CD3^+$) were analyzed in 13 patients diagnosed with visceral leishmaniasis before and after treatment. Compared to healthy controls, there was a greater expression before treatment of TLR2 and TLR4 in lymphocytes and monocytes. Additionally, the levels of TNF-α, IL-10, and TGF-β increased, and those of IFN-γ, IL-17, and NO decreased. Although the expression of these two receptors did not change in lymphocytes after treatment, in monocytes they were found to be lower for TNF-α and IL-10 and higher for TGF-β, IFN-γ, IL-17, and NO [101].

Samples of whole blood taken from patients in Eastern Sudan with visceral leishmaniasis were stimulated with live *L. donovani* promastigotes. The expression of TLR2, TLR4, and TLR9 was found, which correlated with the production of IFN-γ, TNF-α, and IL-10 [102]. Similarly, the expression of TLR2 and TLR4 was measured in peripheral blood mononuclear cells of patients with cutaneous leishmaniasis, both those with healing and non-healing wounds, observing a significantly greater level in the macrophages of individuals with the healing versus non-healing lesions. This suggests a possible role of TLR2 and TLR4 in the outcome of cutaneous leishmaniasis lesions [103].

It is known that *L.* (*V.*) *braziliensis* and *L.* (L.) *amazonensis* interact with these same TLRs to promote a differential T-cell immune response and cytokine expression in mucosal leishmaniasis and anthroponotic cutaneous leishmaniasis. Biopsies taken from skin and mucosal lesions of infected patients were examined by immunohistochemistry, finding an important expression of TLR2, TLR4, and TLR9. Whereas tissues associated with *L.* (*V.*) *braziliensis* exhibited strong expression of TLR2 and TLR4, those tissues linked to *L.* (L.) *amazonensis* displayed similar results in relation to TLR9. The greatest expression of $CD4^+$ T cells was encountered in mucosal leishmaniasis and the lowest in anergic diffuse cutaneous leishmaniasis. Similarly, $CD8^+$ T cells showed their lowest expression in the latter disorder compared to the other forms of the disease. There was greater expression of TNF-α in anergic diffuse cutaneous leishmaniasis versus mucosal leishmaniasis and of IL-10 and TGF-β in mucosal leishmaniasis versus anergic diffuse cutaneous leishmaniasis [4].

Although NK cells are detected in individuals infected with *L. mexicana* and suffering from localized cutaneous leishmaniasis as well as diffuse cutaneous leishmaniasis, the number of cells and the effector mechanisms differed drastically between the two groups. The number of NK cells, production of IFN-γ and TNF-α, and expression of TLR2, TLR1, and TLR6 were all lower than normal in patients with diffuse cutaneous leishmaniasis while being normal in those with localized cutaneous leishmaniasis. The altered protein expression found in NK cells of the former group correlated with the downregulation of IFN-γ gene expression in LPG-stimulated and non-stimulated cells. In conclusion, the lower number of NK cells and their limited activity in individuals with diffuse cutaneous leishmaniasis, evidenced by reduced TLR expression and cytokine production, are likely involved in the severity of the disease [5].

Peripheral blood was taken from patients with anthroponotic cutaneous leishmaniasis, including those responsive and unresponsive to treatment as well as healthy controls. Some mononuclear cells from these blood samples were exposed to *L. tropica* and others unexposed. An evaluation was made of the gene expression of TLR2, TLR4, TLR9, and TNF-α and the activity of iNOS and arginase in monocytes from patients unresponsive and responsive to Glucantime treatment. Upon comparing the monocytes exposed and unexposed to *L. tropica*, the former exhibited greater expression of all three TLRs and TNF-α and lower expression of iNOS in both groups of patients (responsive and unresponsive to treatment). Additionally, there was a significant downregulation of TLR2 and TNF-α expression and upregulation of TLR9 expression in isolates from unresponsive versus responsive individuals. Isolates from the former group also showed a significant increase in the level of arginase in monocytes stimulated with *L. tropica* and cultured promastigotes [6].

7.2 Chronic chagasic cardiomyopathy (CCC)

Chagas disease has different clinical forms: indeterminate, digestive, and cardiodigestive. CCC is associated with greater expression of TLR2, IL-12, and TNF-α. However, the expression of MyD88 mRNA is greater in cardiodigestive than indeterminate and cardiac patients. Serum mRNA expression of IL-12 and TNF-α transcripts was found in cardiac patients, who showed a higher production of TLR-induced inflammatory cytokines (TNF-α and IL-12) than indeterminate patients and uninfected individuals. Digestive and cardiodigestive clinical cases are correlated with elevated mRNA expression of TLR8 and IFN-β [104].

7.3 Helminth infections

During the interaction of *S. mansoni* with the host immune system, TLR4 activation provides a protective role against infection, while TLR2 activation is favorable for the parasite. The gene expression of the TLRs 1, 3, 7, and 8 is suppressed after infection. The antigens of *A. lumbricoides* promote the expression of TLR2 and thus engender a Th2 response [105–108].

8. Role of TLRs in strategies for the control of parasitic infections

TLRs have potential as therapeutic targets, either alone or in combination with conventional immunotherapy and pharmacotherapy. In recent years, antagonists of TLRs or agonists of their negative regulators have been investigated as vaccine adjuvants to enhance an effective immune response against tumors, allergies, and infectious diseases [109].

The vaccine adjuvant properties of TLR7 and/or TLR8 agonists imiquimod and R848 were tested in the model of infection by *L. major*, determining the immune response before and after infection. Protective immunity was generated following subcutaneous but not intramuscular vaccination [110]. In another study, the effect of *L. major* polyclonal anti-murine TLR2 and TLR4 antibodies was assessed on cutaneous leishmaniasis and inflammatory arthritis. Both antibodies suppressed the development of clinical parameters, accompanied by reduced pro-inflammatory cytokine production. Hence, anti-TLR2 and TLR4 antibodies possibly have a synergistic therapeutic effect on inflammatory disease [111].

In visceral leishmaniasis, evaluation was made of anti-*Leishmania* immune responses. The protective efficacy of the glycosphingophospholipid (GSPL) antigen of *L. donovani* parasites when acting as a ligand for β-(1-4)-galactose terminal NKT cells suggests an important role of TLR4. This receptor may function as an upstream sensor by GSPL and induce the intracellular inflammatory signaling necessary for killing parasites. Treatment with GSPL was able to cause a highly effective T-cell response that contributed to good control of infection. Therefore, the synergism of TLR4 and NKT cells prompted GSPL to evoke a host-protective immunological response in experimental visceral leishmaniasis [112].

The collateral effects of a malaria infection during pregnancy correlate with immune activation in placental tissue. Since TLR4 plays a key role in this process, its blockage could be a potential strategy for therapeutic intervention to reduce the incidence of malaria-induced pathology both in the mother and the fetus [113].

Skin scarification with the *P. falciparum* peptide vaccine in combination with a TLR agonist produces systemic neutralizing antibodies with the potential of blocking parasite egress from the skin (and thus avoiding the invasion of liver cells) [114].

A good adjuvant formulation is crucial in the development of a successful vaccine. In the amebiasis model, the nanoliposome adjuvant containing synergistic TLR4 and TLR7/8 agonists successfully elicited balanced systemic humoral and cellular immune responses. The immunization protected against infection with up to 55% efficacy [115].

Since helminths use various molecules to regulate the host immune response, some of these may have potential therapeutic action against allergies and other inflammatory diseases. This anti-inflammatory strategy has been successful in treating diseases in animal models. Helminth-derived molecules are potent immunomodulators that could possibly be used for the design of new anti-inflammatory drugs. For example, the nematodes *T. suis* and *T. spiralis* induce a significant suppression of symptoms in autoimmune encephalomyelitis, an animal model validated for multiple sclerosis. Therefore, infection with live nematodes is not a prerequisite for the suppression of inflammation [53, 116].

Crohn's disease and ulcerative colitis are closely related to inflammatory processes. Some therapies, such as the ingestion of eggs of *T. suis*, promote a TLR2- and TLR4-regulated decline in inflammatory activity as well as a reduction in adverse effects of inflammation. This is based on the ability of helminths to polarize the response of helper T cells to a Th2 type, which inhibits inflammation. *F. hepatica*, on the other hand, exerts influence on dendritic cells activated by CpG, thus fostering the development of Tregs and a decrease in the severity and incidence of the disease [117]. *S. mansoni* alleviates allergies and reduces inflammation in airways. *A. suum* diminishes ocular allergic disease and *T. spiralis*, by evoking a Th2 response, and inhibits the production of IFN-γ to relieve colitis. These results are interesting because they demonstrate the interference of a helminth infection with the establishment of an inflammatory immune response that favors other pathologies [118]. The aforementioned evidence of helminths regulating the immune response through TLRs could be instrumental in the development of therapeutic targets as well as in the inhibition or stimulation of their expression. Further research is needed to clarify the potential role of helminths in the modulation of the inflammatory response.

9. Conclusion

TLRs, one of the best characterized families of receptors, have a critical role in the host defense against infection. Additionally, they play a key role in the capacity of different protozoans and helminths to be able to generate a continuous activation of the host immune system by modulating the elements of the innate and/or adaptive response. Such intervention by these parasites in the immune response is aimed at promoting their survival inside the host. Single-nucleotide polymorphisms in TLRs participate importantly in increasing parasitemia in the host. However, agonists of TLRs can have a dual role. Whereas they may serve as adjuvants or vaccines to promote the maturation of dendritic cells and thus induce an adaptive immune response, they are also capable of triggering inflammatory cytokine production that has a pathogenic role in many diseases. Consequently, antibodies to TLRs and inhibitors of TLR signaling pathways have considerable potential as therapeutic agents. It is still necessary to clarify their mechanisms for modulating the response of these receptors to be able to design and develop innovative therapeutic targets.

Acknowledgements

The authors are grateful to Bruce Allan Larsen for proofreading the manuscript, and also thank Omar A. Garcia Hernández and Ofelia Pérez Olvera for technical assistance. This work was supported by grants 284018 from CONACyT México and IN218619 from DGAPA-PAPIIT to MMAG.

Conflict of interest

The authors declare that the research was conducted in the absence of any commercial or financial relationships that could be construed as a potential conflict of interest.

Author details

M. Magdalena Aguirre-García[1*], Araceli Rojas-Bernabé[2,3], A. Pamela Gómez-García[1] and Alma R. Escalona-Montaño[1]

1 División de Investigación, Facultad de Medicina, Instituto Nacional de Cardiología Ignacio Chávez, Unidad de Investigación UNAM-INC, Ciudad de México, Mexico

2 Escuela Superior de Medicina, Instituto Politécnico Nacional, Salvador Díaz Mirón, Col. Casco Santo Tomas, Ciudad de México, Mexico

3 Hospital de la Mujer, SSA, Col. Santo Tomas, Ciudad de México, Mexico

*Address all correspondence to: maguirre@unam.mx

References

[1] Achtman JC, Ellis DL, Saylors B, Boh EE. Cutaneous leishmaniasis caused by Leishmania (Viannia) panamensis in 2 travelers. JAAD Case Reports. 2016;**2**(2):95-97. Available from: http://www.ncbi.nlm.nih.gov/pubmed/27051841

[2] Sacramento L, Trevelin SC, Nascimento MS, Lima-Junior DS, Costa DL, Almeida RP, et al. Toll-like receptor 9 signaling in dendritic cells regulates neutrophil recruitment to inflammatory foci following *Leishmania infantum* infection. Infection and Immunity. 2015;**83**(12):4604-4616. Available from: http://www.ncbi.nlm.nih.gov/pubmed/26371124

[3] Cezario GA, de Oliveira LR, Peresi E, Nicolete VC, Polettini J, de Lima CR, et al. Analysis of the expression of toll-like receptors 2 and 4 and cytokine production during experimental *Leishmania chagasi* infection. Memórias do Instituto Oswaldo Cruz. 2011;**106**(5):573-583. Available from: http://www.ncbi.nlm.nih.gov/pubmed/21894379

[4] Campos MB, Lima L, de Lima ACS, Vasconcelos Dos Santos T, Ramos PKS, Gomes CMC, et al. Toll-like receptors 2, 4, and 9 expressions over the entire clinical and immunopathological spectrum of American cutaneous leishmaniasis due to *Leishmania* (V.) *braziliensis* and *Leishmania* (L.) *amazonensis*. PLoS One. 2018;**13**(3):e0194383. Available from: http://www.ncbi.nlm.nih.gov/pubmed/29543867

[5] Caneda-Guzman IC, Salaiza-Suazo N, Fernandez-Figueroa EA, Carrada-Figueroa G, Aguirre-Garcia M, Becker I. NK cell activity differs between patients with localized and diffuse cutaneous leishmaniasis infected with *Leishmania mexicana*: A comparative study of TLRs and cytokines. PLoS One. 2014;**9**(11):e112410. Available from: http://www.ncbi.nlm.nih.gov/pubmed/25397678

[6] Oliaee RT, Sharifi I, Afgar A, Jafarzadeh A, Kareshk AT, Bamorovat M, et al. Differential expression of TLRs 2, 4, 9, iNOS and TNF-alpha and arginase activity in peripheral blood monocytes from glucantime unresponsive and responsive patients with anthroponotic cutaneous leishmaniasis caused by *Leishmania tropica*. Microbial Pathogenesis. 2018;**126**:368-378. Available from: http://www.ncbi.nlm.nih.gov/pubmed/30399441

[7] Teixeira ARL, Nascimento RJ, Sturm NR. Evolution and pathology in chagas disease—A review. Memórias do Instituto Oswaldo Cruz. 2006;**101**(5):463-491. Available from: http://www.ncbi.nlm.nih.gov/pubmed/17072450

[8] Dutra WO, Gollob KJ. Current concepts in immunoregulation and pathology of human Chagas disease. Current Opinion in Infectious Diseases. 2008;**21**(3):287-292. Available from: http://www.ncbi.nlm.nih.gov/pubmed/18448974

[9] Grace E, Asbill S, Virga K. Naegleria fowleri: Pathogenesis, diagnosis, and treatment options. Antimicrobial Agents and Chemotherapy. 2015;**59**(11):6677-6681. Available from: http://www.ncbi.nlm.nih.gov/pubmed/26259797

[10] Khan NA. Acanthamoeba: Biology and increasing importance in human health. FEMS Microbiology Reviews. 2006;**30**(4):564-595. Available from: http://www.ncbi.nlm.nih.gov/pubmed/16774587

[11] Martinez AJ, Visvesvara GS. Free-living, amphizoic and opportunistic amebas. Brain Pathology.

1997;7(1):583-598. Available from: http://www.ncbi.nlm.nih.gov/pubmed/9034567

[12] Adam RD. Biology of *Giardia lamblia*. Clinical Microbiology Reviews. 2001;**14**(3):447-475. Available from: http://www.ncbi.nlm.nih.gov/pubmed/11432808

[13] Schofield L, Grau GE. Immunological processes in malaria pathogenesis. Nature Reviews. Immunology. 2005;**5**(9):722-735. Available from: http://www.ncbi.nlm.nih.gov/pubmed/16138104

[14] Petrin D, Delgaty K, Bhatt R, Garber G. Clinical and microbiological aspects of *Trichomonas vaginalis*. Clinical Microbiology Reviews. 1998;**11**(2):300-317. Available from: http://www.ncbi.nlm.nih.gov/pubmed/9564565

[15] Kotloff KL, Nataro JP, Blackwelder WC, Nasrin D, Farag TH, Panchalingam S, et al. Burden and aetiology of diarrhoeal disease in infants and young children in developing countries (the Global Enteric Multicenter Study, GEMS): A prospective, case-control study. Lancet. 2013;**382**(9888):209-222. Available from: http://www.ncbi.nlm.nih.gov/pubmed/23680352

[16] Arisue N, Hashimoto T, Yoshikawa H, Nakamura Y, Nakamura G, Nakamura F, et al. Phylogenetic position of *Blastocystis* hominis and of stramenopiles inferred from multiple molecular sequence data. The Journal of Eukaryotic Microbiology. 2002;**49**(1):42-53. Available from: http://www.ncbi.nlm.nih.gov/pubmed/11908898

[17] Tan KS, Mirza H, Teo JD, Wu B, Macary PA. Current views on the clinical relevance of *Blastocystis* spp. Current Infectious Disease Reports. 2010;**12**(1):28-35. Available from: http://www.ncbi.nlm.nih.gov/pubmed/21308496

[18] Akira S, Takeda K, Kaisho T. Toll-like receptors: Critical proteins linking innate and acquired immunity. Nature Immunology. 2001;**2**(8):675-680. Available from: http://www.nature.com/articles/ni0801_675

[19] Dissanayake S, Amith RS, Shahin A. Taenia crassiceps carbohydrates stimulate IL-6 expression in naïve murine macrophages via Toll-like receptors (TLRs). Molecular Immunology. 2004;**41**(4):391-398. Available from: http://linkinghub.elsevier.com/retrieve/pii/S0161589004000835

[20] Du L, Liu L, Yu Y, Shan H, Li L. Trichinella spiralis excretory-secretory products protect against polymicrobial sepsis by suppressing MyD88 via mannose receptor. BioMed Research International. 2014;**2014**:898646. Available from: https://www.hindawi.com/journals/bmri/2014/898646/

[21] Hawn TR, Ozinsky A, Underhill DM, Buckner FS, Akira S, Aderem A. *Leishmania major* activates IL-1 alpha expression in macrophages through a MyD88-dependent pathway. Microbes and Infection. 2002;**4**(8):763-771. Available from: http://www.ncbi.nlm.nih.gov/pubmed/12270723

[22] Becker I, Salaiza N, Aguirre M, Delgado J, Carrillo-Carrasco N, Kobeh LG, et al. Leishmania lipophosphoglycan (LPG) activates NK cells through toll-like receptor-2. Molecular and Biochemical Parasitology. 2003;**130**(2):65-74. Available from: http://www.ncbi.nlm.nih.gov/pubmed/12946842

[23] Galdino H Jr, Saar Gomes R, Dos Santos JC, Pessoni LL, Maldaner AE, Marques SM, et al. Leishmania (Viannia) braziliensis amastigotes induces the expression of TNFalpha and IL-10 by human peripheral blood mononuclear cells in vitro in a

TLR4-dependent manner. Cytokine. 2016;**88**:184-192. Available from: http://www.ncbi.nlm.nih.gov/pubmed/27649507

[24] Schleicher U, Liese J, Knippertz I, Kurzmann C, Hesse A, Heit A, et al. NK cell activation in visceral leishmaniasis requires TLR9, myeloid DCs, and IL-12, but is independent of plasmacytoid DCs. The Journal of Experimental Medicine. 2007;**204**(4):893-906. Available from: http://www.ncbi.nlm.nih.gov/pubmed/17389237

[25] Martínez-Salazar B, Berzunza-Cruz M, Becker I. *Leishmania mexicana* DNA activates murine macrophages and increases their TLR9 expression. Gaceta Médica de México. 2008;**144**(2):99-104. Available from: http://www.ncbi.nlm.nih.gov/pubmed/18590029

[26] Abou Fakher FH, Rachinel N, Klimczak M, Louis J, Doyen N. TLR9-dependent activation of dendritic cells by DNA from *Leishmania major* favors Th1 cell development and the resolution of lesions. Journal of Immunology. 2009;**182**(3):1386-1396. Available from: http://www.ncbi.nlm.nih.gov/pubmed/19155485

[27] Srivastava S, Pandey SP, Jha MK, Chandel HS, Saha B. Leishmania expressed lipophosphoglycan interacts with Toll-like receptor (TLR)-2 to decrease TLR-9 expression and reduce anti-leishmanial responses. Clinical and Experimental Immunology. 2013;**172**(3):403-409. Available from: http://www.ncbi.nlm.nih.gov/pubmed/23600828

[28] Gallego C, Golenbock D, Gomez MA, Saravia NG. Toll-like receptors participate in macrophage activation and intracellular control of Leishmania (Viannia) panamensis. Infection and Immunity. 2011;**79**(7):2871-2879. Available from: http://iai.asm.org/lookup/doi/10.1128/IAI.01388-10

[29] Rojas-Bernabe A, Garcia-Hernandez O, Maldonado-Bernal C, Delegado-Dominguez J, Ortega E, Gutierrez-Kobeh L, et al. *Leishmania mexicana* lipophosphoglycan activates ERK and p38 MAP kinase and induces production of proinflammatory cytokines in human macrophages through TLR2 and TLR4. Parasitology. 2014;**141**(6):788-800. Available from: http://www.ncbi.nlm.nih.gov/pubmed/24512642

[30] Campos MA, Almeida IC, Takeuchi O, Akira S, Valente EP, Procopio DO, et al. Activation of toll-like receptor-2 by glycosylphosphatidylinositol anchors from a protozoan parasite. Journal of Immunology. 2001;**167**(1):416-423. Available from: http://www.ncbi.nlm.nih.gov/pubmed/11418678

[31] Oliveira AC, Peixoto JR, de Arruda LB, Campos MA, Gazzinelli RT, Golenbock DT, et al. Expression of functional TLR4 confers proinflammatory responsiveness to *Trypanosoma cruzi* glycoinositolphospholipids and higher resistance to infection with *T. cruzi*. Journal of Immunology. 2004;**173**(9):5688-5696. Available from: http://www.ncbi.nlm.nih.gov/pubmed/15494520

[32] Bott E, Carneiro AB, Gimenez G, Lopez MG, Lammel EM, Atella GC, et al. Lipids from *Trypanosoma cruzi* Amastigotes of RA and K98 strains generate a pro-inflammatory response via TLR2/6. Frontiers in Cellular and Infection Microbiology. 2018;**8**:151. Available from: http://www.ncbi.nlm.nih.gov/pubmed/29868507

[33] Hisaeda H, Tetsutani K, Imai T, Moriya C, Tu L, Hamano S, et al. Malaria parasites require TLR9 signaling for immune evasion by activating regulatory T cells. Journal of Immunology. 2008;**180**(4):2496-2503. Available from: http://www.ncbi.nlm.nih.gov/pubmed/18250459

[34] Coban C, Ishii KJ, Kawai T, Hemmi H, Sato S, Uematsu S, et al. Toll-like receptor 9 mediates innate immune activation by the malaria pigment hemozoin. The Journal of Experimental Medicine. 2005;**201**(1):19-25. Available from: http://www.ncbi.nlm.nih.gov/pubmed/15630134

[35] Li L, Li X, Gong P, Zhang X, Yang Z, Yang J, et al. Trichomonas vaginalis induces production of proinflammatory cytokines in mouse macrophages through activation of MAPK and NF-kappaB pathways partially mediated by TLR2. Frontiers in Microbiology. 2018;**9**:712. Available from: http://www.ncbi.nlm.nih.gov/pubmed/29692771

[36] Riestra AM, Valderrama JA, Patras KA, Booth SD, Quek XY, Tsai CM, et al. *Trichomonas vaginalis* induces NLRP3 inflammasome activation and pyroptotic cell death in human macrophages. Journal of Innate Immunity. 2019;**11**(1):86-98. Available from: http://www.ncbi.nlm.nih.gov/pubmed/30391945

[37] Im SJ, Han IH, Kim JH, Gu NY, Seo MY, Chung YH, et al. Inflammatory response of a prostate stromal cell line induced by *Trichomonas vaginalis*. Parasite Immunology. 2016;**38**(4): 218-227. Available from: http://www.ncbi.nlm.nih.gov/pubmed/26832322

[38] Yarovinsky F, Zhang D, Andersen JF, Bannenberg GL, Serhan CN, Hayden MS, et al. TLR11 activation of dendritic cells by a protozoan profilin-like protein. Science. 2005;**308**(5728):1626-1629. Available from: http://www.sciencemag.org/cgi/doi/10.1126/science.1109893

[39] Maldonado-Bernal C, Kirschning CJ, Rosenstein Y, Rocha LM, Rios-Sarabia N, Espinosa-Cantellano M, et al. The innate immune response to *Entamoeba histolytica* lipopeptidophosphoglycan is mediated by toll-like receptors 2 and 4. Parasite Immunology. 2005;**27**(4):127-137. Available from: http://doi.wiley.com/10.1111/j.1365-3024.2005.00754.x

[40] Ivory CP, Prystajecky M, Jobin C, Chadee K. Toll-like receptor 9-dependent macrophage activation by *Entamoeba histolytica* DNA. Infection and Immunity. 2008;**76**(1):289-297. Available from: http://www.ncbi.nlm.nih.gov/pubmed/17984204

[41] Lim EJ, Lee SH, Lee JG, Kim JR, Yun SS, Baek SH, et al. Toll-like receptor 9 dependent activation of MAPK and NF-kB is required for the CpG ODN-induced matrix metalloproteinase-9 expression. Experimental & Molecular Medicine. 2007;**39**(2):239-245. Available from: http://www.ncbi.nlm.nih.gov/pubmed/17464186

[42] Li X, Zhang X, Gong P, Xia F, Li L, Yang Z, et al. TLR2(−/−) mice display decreased severity of giardiasis via enhanced proinflammatory cytokines production dependent on AKT signal pathway. Frontiers in Immunology. 2017;**8**:1186. Available from: http://www.ncbi.nlm.nih.gov/pubmed/28979269

[43] Teo JDW, Macary PA, Tan KSW. Pleiotropic effects of *Blastocystis* spp. subtypes 4 and 7 on ligand-specific toll-like receptor signaling and NF-κB activation in a human monocyte cell line. PLoS One. 2014;**9**(2):e89036. Available from: https://dx.plos.org/10.1371/journal.pone.0089036

[44] Martínez-Castillo M, Santos-Argumedo L, Galván-Moroyoqui JM, Serrano-Luna J, Shibayama M. Toll-like receptors participate in *Naegleria fowleri* recognition. Parasitology Research. 2018;**117**(1):75-87. Available from: http://link.springer.com/10.1007/s00436-017-5666-9

[45] Aksoy E, Zouain CS, Vanhoutte F, Fontaine J, Pavelka N, Thieblemont N, et al. Double-stranded RNAs from the

helminth parasite Schistosoma activate TLR3 in dendritic cells. The Journal of Biological Chemistry. 2005;**280**(1): 277-283. Available from: http://www.jbc.org/lookup/doi/10.1074/jbc.M411223200

[46] Aranzamendi C, Sofronic-Milosavljevic L, Pinelli E. Helminths: Immunoregulation and inflammatory diseases-which side are Trichinella spp. and Toxocara spp. on? Journal of Parasitology Research. 2013;**2013**:329438. Available from: http://www.hindawi.com/journals/jpr/2013/329438/

[47] Babu S, Blauvelt CP, Kumaraswami V, Nutman TB. Diminished expression and function of TLR in lymphatic filariasis: A novel mechanism of immune dysregulation. Journal of Immunology. 2005;**175**(2):1170-1176. Available from: http://www.ncbi.nlm.nih.gov/pubmed/16002719

[48] Carranza F, Falcón CR, Nuñez N, Knubel C, Correa SG, Bianco I, et al. Helminth antigens enable CpG-activated dendritic cells to inhibit the symptoms of collagen-induced arthritis through Foxp3+ regulatory T cells. PLoS One. 2012;**7**(7):e40356. Available from: https://dx.plos.org/10.1371/journal.pone.0040356

[49] Cervi L, MacDonald AS, Kane C, Dzierszinski F, Pearce EJ. Cutting edge: Dendritic cells copulsed with microbial and helminth antigens undergo modified maturation, segregate the antigens to distinct intracellular compartments, and concurrently induce microbe-specific Th1 and helminth-specific Th2 responses. Journal of Immunology. 2004;**172**(4):2016-2020. Available from: http://www.ncbi.nlm.nih.gov/pubmed/14764665

[50] Chen C-C, Louie S, McCormick B, Walker WA, Shi HN. Concurrent infection with an intestinal helminth parasite impairs host resistance to enteric *Citrobacter rodentium* and enhances Citrobacter-induced colitis in mice. Infection and Immunity. 2005;**73**(9):5468-5481. Available from: http://iai.asm.org/cgi/doi/10.1128/IAI.73.9.5468-5481.2005

[51] Coakley G, Buck AH, Maizels RM. Host parasite communications-messages from helminths for the immune system: Parasite communication and cell-cell interactions. Molecular and Biochemical Parasitology. 2016;**208**(1):33-40. Available from: https://linkinghub.elsevier.com/retrieve/pii/S0166685116300755

[52] Correale J, Farez M. Helminth antigens modulate immune responses in cells from multiple sclerosis patients through TLR2-dependent mechanisms. Journal of Immunology. 2009;**183**(9):5999-6012. Available from: http://www.ncbi.nlm.nih.gov/pubmed/19812189

[53] Donnelly S, O'Neill SM, Stack CM, Robinson MW, Turnbull L, Whitchurch C, et al. Helminth cysteine proteases inhibit TRIF-dependent activation of macrophages via degradation of TLR3. The Journal of Biological Chemistry. 2010;**285**(5):3383-3392. Available from: http://www.jbc.org/lookup/doi/10.1074/jbc.M109.060368

[54] Everts B, Smits HH, Hokke CH, Yazdanbakhsh M. Helminths and dendritic cells: Sensing and regulating via pattern recognition receptors, Th2 and Treg responses. European Journal of Immunology. 2010;**40**(6): 1525-1537. Available from: http://doi.wiley.com/10.1002/eji.200940109

[55] Falcón C, Carranza F, Martínez FF, Knubel CP, Masih DT, Motrán CC, et al. Excretory-secretory products (ESP) from *Fasciola hepatica* induce tolerogenic properties in myeloid dendritic cells. Veterinary Immunology and Immunopathology. 2010;**137** (1-2):36-46. Available from:

https://linkinghub.elsevier.com/retrieve/pii/S0165242710001224

[56] Gao Y, Zhang M, Chen L, Hou M, Ji M, Wu G. Deficiency in TLR2 but not in TLR4 impairs dendritic cells derived IL-10 responses to schistosome antigens. Cellular Immunology. 2012;**272**(2):242-250. Available from: http://www.ncbi.nlm.nih.gov/pubmed/22063738

[57] Giacomin P, Croese J, Krause L, Loukas A, Cantacessi C. Suppression of inflammation by helminths: A role for the gut microbiota? Philosophical Transactions of the Royal Society B: Biological Sciences. 2015;**370**(1675):20140296. Available from: http://www.ncbi.nlm.nih.gov/pubmed/26150662

[58] Goodridge HS, Marshall FA, Else KJ, Houston KM, Egan C, Al-Riyami L, et al. Immunomodulation via novel use of TLR4 by the filarial nematode phosphorylcholine-containing secreted product, ES-62. Journal of Immunology. 2005;**174**(1):284-293. Available from: http://www.ncbi.nlm.nih.gov/pubmed/15611251

[59] Harnett W, Harnett MM. Therapeutic immunomodulators from nematode parasites. Expert Reviews in Molecular Medicine. 2008;**10**:e18. Available from: http://www.ncbi.nlm.nih.gov/pubmed/18561863

[60] Haslam SM, Morris HR, Dell A. Mass spectrometric strategies: Providing structural clues for helminth glycoproteins. Trends in Parasitology. 2001;**17**(5):231-235. Available from: http://www.ncbi.nlm.nih.gov/pubmed/11323307

[61] Hurst RJM, Else KJ. *Trichuris muris* research revisited: A journey through time. Parasitology. 2013;**140**(11):1325-1339. Available from: http://www.journals.cambridge.org/abstract_S0031182013001054

[62] Johnston MJG, MacDonald JA, McKay DM. Parasitic helminths: A pharmacopeia of anti-inflammatory molecules. Parasitology. 2009;**136**(2):125-147. Available from: http://www.journals.cambridge.org/abstract_S0031182008005210

[63] Kuijk LM, Klaver EJ, Kooij G, van der Pol SMA, Heijnen P, Bruijns SCM, et al. Soluble helminth products suppress clinical signs in murine experimental autoimmune encephalomyelitis and differentially modulate human dendritic cell activation. Molecular Immunology. 2012;**51**(2):210-218. Available from: https://linkinghub.elsevier.com/retrieve/pii/S0161589012001964

[64] Ludwig-Portugall I, Layland LE. TLRs, Treg, and B cells, an interplay of regulation during helminth infection. Frontiers in Immunology. 2012;**3**:8. Available from: http://journal.frontiersin.org/article/10.3389/fimmu.2012.00008/abstract

[65] Kropf P, Freudenberg MA, Modolell M, Price HP, Herath S, Antoniazi S, et al. Toll-like receptor 4 contributes to efficient control of infection with the protozoan parasite *Leishmania major*. Infection and Immunity. 2004;**72**(4):1920-1928. Available from: http://www.ncbi.nlm.nih.gov/pubmed/15039311

[66] Kropf P, Freudenberg N, Kalis C, Modolell M, Herath S, Galanos C, et al. Infection of C57BL/10ScCr and C57BL/10ScNCr mice with *Leishmania major* reveals a role for toll-like receptor 4 in the control of parasite replication. Journal of Leukocyte Biology. 2004;**76**(1):48-57. Available from: http://doi.wiley.com/10.1189/jlb.1003484

[67] Pandey SP, Chandel HS, Srivastava S, Selvaraj S, Jha MK, Shukla D, et al. Pegylated bisacycloxypropylcysteine, a diacylated lipopeptide ligand of TLR6, plays a host-protective role

against experimental *Leishmania major* infection. Journal of Immunology. 2014;**193**(7):3632-3643. Available from: http://www.ncbi.nlm.nih.gov/pubmed/25194056

[68] Shukla D, Chandel HS, Srivastava S, Chauhan P, Pandey SP, Patidar A, et al. TLR11 or TLR12 silencing reduces *Leishmania major* infection. Cytokine. 2018;**104**:110-113. Available from: http://www.ncbi.nlm.nih.gov/pubmed/29017773

[69] Vargas-Inchaustegui DA, Tai W, Xin L, Hogg AE, Corry DB, Soong L. Distinct roles for MyD88 and toll-like receptor 2 during *Leishmania braziliensis* infection in mice. Infection and Immunity. 2009;**77**(7):2948-2956. Available from: http://iai.asm.org/cgi/doi/10.1128/IAI.00154-09

[70] Vieira ÉLM, Keesen TSL, Machado PR, Guimarães LH, Carvalho EM, Dutra WO, et al. Immunoregulatory profile of monocytes from cutaneous leishmaniasis patients and association with lesion size. Parasite Immunology. 2013;**35**(2):65-72. Available from: http://doi.wiley.com/10.1111/pim.12012

[71] Srivastav S, Kar S, Chande AG, Mukhopadhyaya R, Das PK. *Leishmania donovani* exploits host deubiquitinating enzyme A20, a negative regulator of TLR signaling, to subvert host immune response. Journal of Immunology. 2012;**189**(2):924-934. Available from: http://www.ncbi.nlm.nih.gov/pubmed/22685311

[72] Basu M, Maji AK, Chakraborty A, Banerjee R, Mullick S, Saha P, et al. Genetic association of toll-like-receptor 4 and tumor necrosis factor-alpha polymorphisms with *Plasmodium falciparum* blood infection levels. Infection, Genetics and Evolution. 2010;**10**(5):686-696. Available from: https://linkinghub.elsevier.com/retrieve/pii/S1567134810000687

[73] Hamann L, Bedu-Addo G, Eggelte TA, Schumann RR, Mockenhaupt FP. The toll-like receptor 1 variant S248N influences placental malaria. Infection, Genetics and Evolution. 2010;**10**(6):785-789. Available from: http://www.ncbi.nlm.nih.gov/pubmed/20478407

[74] Ferwerda B, MB MC, Alonso S, Giamarellos-Bourboulis EJ, Mouktaroudi M, Izagirre N, et al. TLR4 polymorphisms, infectious diseases, and evolutionary pressure during migration of modern humans. Proceedings of the National Academy of Sciences of the United States of America. 2007;**104**(42):16645-16650. Available from: http://www.ncbi.nlm.nih.gov/pubmed/17925445

[75] Leoratti FM, Farias L, Alves FP, Suarez-Mutis MC, Coura JR, Kalil J, et al. Variants in the toll-like receptor signaling pathway and clinical outcomes of malaria. The Journal of Infectious Diseases. 2008;**198**(5):772-780. Available from: http://www.ncbi.nlm.nih.gov/pubmed/18662133

[76] Khor CC, Chapman SJ, Vannberg FO, Dunne A, Murphy C, Ling EY, et al. A Mal functional variant is associated with protection against invasive pneumococcal disease, bacteremia, malaria and tuberculosis. Nature Genetics. 2007;**39**(4):523-528. Available from: http://www.nature.com/articles/ng1976

[77] Rani A, Nawaz SK, Arshad M, Irfan S. Role of rs4986790 polymorphism of TLR4 gene in susceptibility towards malaria infection in the Pakistani population. Iranian Journal of Public Health. 2018;**47**(5):735-741. Available from: http://www.ncbi.nlm.nih.gov/pubmed/29922617

[78] Costa AG, Ramasawmy R, Val FFA, Ibiapina HNS, Oliveira AC, Tarragô AM, et al. Polymorphisms in TLRs influence circulating cytokines production in *Plasmodium vivax* malaria. Cytokine.

2018;**110**:374-380. Available from: http://www.ncbi.nlm.nih.gov/pubmed/29656958

[79] Cramer JP, Lepenies B, Kamena F, Hölscher C, Freudenberg MA, Burchard GD, et al. MyD88/IL-18-dependent pathways rather than TLRs control early parasitaemia in non-lethal *Plasmodium yoelii* infection. Microbes and Infection. 2008;**10**(12-13):1259-1265. Available from: http://linkinghub.elsevier.com/retrieve/pii/S1286457908002190

[80] Caetano BC, Carmo BB, Melo MB, Cerny A, dos Santos SL, Bartholomeu DC, et al. Requirement of UNC93B1 reveals a critical role for TLR7 in host resistance to primary infection with *Trypanosoma cruzi*. Journal of Immunology. 2011;**187**(4):1903-1911. Available from: http://www.jimmunol.org/cgi/doi/10.4049/jimmunol.1003911

[81] Costa AG, Ramasawmy R, Ibiapina HNS, Sampaio VS, Xabregas LA, Brasil LW, et al. Association of TLR variants with susceptibility to *Plasmodium vivax* malaria and parasitemia in the Amazon region of Brazil. PLoS One. 2017;**12**(8):e0183840. Available from: http://www.ncbi.nlm.nih.gov/pubmed/28850598

[82] Melo MB, Kasperkovitz P, Cerny A, Konen-Waisman S, Kurt-Jones EA, Lien E, et al. UNC93B1 mediates host resistance to infection with *Toxoplasma gondii*. PLoS Pathogens. 2010;**6**(8):e1001071. Available from: http://www.ncbi.nlm.nih.gov/pubmed/20865117

[83] McKay DM. The therapeutic helminth? Trends in Parasitology. 2009;**25**(3):109-114. Available from: http://www.ncbi.nlm.nih.gov/pubmed/19167926

[84] McSorley HJ, Maizels RM. Helminth infections and host immune regulation. Clinical Microbiology Reviews. 2012;**25**(4):585-608. Available from: http://www.ncbi.nlm.nih.gov/pubmed/23034321

[85] Mowat AM. Does TLR2 regulate intestinal inflammation? European Journal of Immunology. 2010;**40**(2):318-320. Available from: http://www.ncbi.nlm.nih.gov/pubmed/20039306

[86] Kane CM, Cervi L, Sun J, McKee AS, Masek KS, Shapira S, et al. Helminth antigens modulate TLR-initiated dendritic cell activation. Journal of Immunology. 2004;**173**(12):7454-7461. Available from: http://www.ncbi.nlm.nih.gov/pubmed/15585871

[87] Panther E, Idzko M, Corinti S, Ferrari D, Herouy Y, Mockenhaupt M, et al. The influence of lysophosphatidic acid on the functions of human dendritic cells. Journal of Immunology. 2002;**169**(8):4129-4135. Available from: http://www.ncbi.nlm.nih.gov/pubmed/12370341

[88] Pasare C, Medzhitov R. Toll pathway-dependent blockade of CD4+CD25+ T cell-mediated suppression by dendritic cells. Science. 2003;**299**(5609):1033-1036. Available from: http://www.sciencemag.org/cgi/doi/10.1126/science.1078231

[89] Chandra D, Naik S. *Leishmania donovani* infection down-regulates TLR2-stimulated IL-12p40 and activates IL-10 in cells of macrophage/monocytic lineage by modulating MAPK pathways through a contact-dependent mechanism. Clinical and Experimental Immunology. 2008;**154**(2):224-234. Available from: http://doi.wiley.com/10.1111/j.1365-2249.2008.03741.x

[90] Pinelli E, Brandes S, Dormans J, Fonville M, Hamilton CM, van der Giessen J. *Toxocara canis*: Effect of inoculum size on pulmonary pathology and cytokine expression in BALB/c mice. Experimental Parasitology. 2007;**115**(1):76-82. Available from: http://linkinghub.elsevier.com/retrieve/pii/S0014489406001433

[91] Rodríguez E, Noya V, Cervi L, Chiribao ML, Brossard N, Chiale C, et al. Glycans from *Fasciola hepatica* modulate the host immune response and TLR-induced maturation of dendritic cells. PLoS Neglected Tropical Diseases. 2015;**9**(12):e0004234. Available from: https://dx.plos.org/10.1371/journal.pntd.0004234

[92] Segura M, Su Z, Piccirillo C, Stevenson MM. Impairment of dendritic cell function by excretory-secretory products: A potential mechanism for nematode-induced immunosuppression. European Journal of Immunology. 2007;**37**(7):1887-1904. Available from: http://doi.wiley.com/10.1002/eji.200636553

[93] Shepherd C, Navarro S, Wangchuk P, Wilson D, Daly NL, Loukas A. Identifying the immunomodulatory components of helminths. Parasite Immunology. 2015;**37**(6):293-303. Available from: http://doi.wiley.com/10.1111/pim.12192

[94] Sun S, Wang X, Wu X, Zhao Y, Wang F, Liu X, et al. Toll-like receptor activation by helminths or helminth products to alleviate inflammatory bowel disease. Parasites & Vectors. 2011;**4**(1):186. Available from: http://parasitesandvectors.biomedcentral.com/articles/10.1186/1756-3305-4-186

[95] Terrazas CA, Alcántara-Hernández M, Bonifaz L, Terrazas LI, Satoskar AR. Helminth-excreted/secreted products are recognized by multiple receptors on DCs to block the TLR response and bias Th2 polarization in a cRAF dependent pathway. The FASEB Journal. 2013;**27**(11):4547-4560. Available from: http://www.fasebj.org/doi/10.1096/fj.13-228932

[96] van Liempt E, van Vliet SJ, Engering A, García Vallejo JJ, Bank CMC, Sanchez-Hernandez M, et al. *Schistosoma mansoni* soluble egg antigens are internalized by human dendritic cells through multiple C-type lectins and suppress TLR-induced dendritic cell activation. Molecular Immunology. 2007;**44**(10):2605-2615. Available from: http://linkinghub.elsevier.com/retrieve/pii/S0161589006007310

[97] da Costa TA, Silva MV, Mendes MT, Carvalho-Costa TM, Batista LR, Lages-Silva E, et al. Immunomodulation by *Trypanosoma cruzi*: Toward understanding the association of dendritic cells with infecting TcI and TcII populations. Journal of Immunology Research. 2014;**2014**:962047. Available from: http://www.hindawi.com/journals/jir/2014/962047/

[98] Wojtkowiak-Giera A, Derda M, Kolasa-Wołosiuk A, Hadaś E, Kosik-Bogacka D, Solarczyk P, et al. Toll-like receptors in the brain of mice following infection with Acanthamoeba spp. Parasitology Research. 2016;**115**(11):4335-4344. Available from: http://www.ncbi.nlm.nih.gov/pubmed/27511368

[99] Wojtkowiak-Giera A, Derda M, Kosik-Bogacka D, Kolasa-Wołosiuk A, Solarczyk P, Cholewiński M, et al. Influence of *Artemisia annua* L. on toll-like receptor expression in brain of mice infected with Acanthamoeba sp. Experimental Parasitology. 2018;**185**:17-22. Available from: https://linkinghub.elsevier.com/retrieve/pii/S001448941730365X

[100] Kumar R, Singh OP, Gautam S, Nylen S, Sundar S. Enhanced expression of toll-like receptors 2 and 4, but not 9, in spleen tissue from patients with visceral leishmaniasis. Parasite Immunology. 2014;**36**(12):721-725. Available from: http://www.ncbi.nlm.nih.gov/pubmed/25244363

[101] Gatto M, de Abreu MM, Tasca KI, de Assis Golim M, da Silva LDM, Simão JC, et al. The involvement of TLR2 and TLR4 in cytokine and nitric oxide production in visceral leishmaniasis

patients before and after treatment with anti-leishmanial drugs. PLoS One. 2015;**10**(2):e0117977. Available from: http://www.ncbi.nlm.nih.gov/pubmed/25706930

[102] Babiker DT, Bakhiet SM, Mukhtar MM. *Leishmania donovani* influenced cytokines and toll-like receptors expression among Sudanese visceral leishmaniasis patients. Parasite Immunology. 2015;**37**(8):417-425. Available from: http://www.ncbi.nlm.nih.gov/pubmed/25982946

[103] Tolouei S, Hejazi SH, Ghaedi K, Khamesipour A, Hasheminia SJ. TLR2 and TLR4 in cutaneous leishmaniasis caused by *Leishmania major*. Scandinavian Journal of Immunology. 2013;**78**(5):478-484. Available from: http://doi.wiley.com/10.1111/sji.12105

[104] Pereira NS, Queiroga TBD, Nunes DF, Andrade CM, Nascimento MSL, Do-Valle-Matta MA, et al. Innate immune receptors over expression correlate with chronic chagasic cardiomyopathy and digestive damage in patients. PLoS Neglected Tropical Diseases. 2018;**12**(7):e0006589. Available from: https://dx.plos.org/10.1371/journal.pntd.0006589

[105] van Riet E, Everts B, Retra K, Phylipsen M, van Hellemond JJ, Tielens AGM, et al. Combined TLR2 and TLR4 ligation in the context of bacterial or helminth extracts in human monocyte derived dendritic cells: Molecular correlates for Th1/Th2 polarization. BMC Immunology. 2009;**10**(1):9. Available from: http://bmcimmunol.biomedcentral.com/articles/10.1186/1471-2172-10-9

[106] Venugopal PG, Nutman TB, Semnani RT. Activation and regulation of toll-like receptors (TLRs) by helminth parasites. Immunologic Research. 2009;**43**(1-3):252-263. Available from: http://link.springer.com/10.1007/s12026-008-8079-0

[107] Yu Y-R, Deng M-J, Lu W-W, Jia M-Z, Wu W, Qi Y-F. Systemic cytokine profiles and splenic toll-like receptor expression during *Trichinella spiralis* infection. Experimental Parasitology. 2013;**134**(1):92-101. Available from: https://linkinghub.elsevier.com/retrieve/pii/S0014489413000568

[108] Zaccone P, Fehérvári Z, Jones FM, Sidobre S, Kronenberg M, Dunne DW, et al. *Schistosoma mansoni* antigens modulate the activity of the innate immune response and prevent onset of type 1 diabetes. European Journal of Immunology. 2003;**33**(5):1439-1449. Available from: http://doi.wiley.com/10.1002/eji.200323910

[109] Kanzler H, Barrat FJ, Hessel EM, Coffman RL. Therapeutic targeting of innate immunity with toll-like receptor agonists and antagonists. Nature Medicine. 2007;**13**(5):552-559. Available from: http://www.ncbi.nlm.nih.gov/pubmed/17479101

[110] Zhang W-W, Matlashewski G. Immunization with a toll-like receptor 7 and/or 8 agonist vaccine adjuvant increases protective immunity against *Leishmania major* in BALB/c mice. Infection and Immunity. 2008;**76**(8):3777-3783. Available from: http://www.ncbi.nlm.nih.gov/pubmed/18474642

[111] Komai-Koma M, Li D, Wang E, Vaughan D, Xu D. Anti-toll-like receptor 2 and 4 antibodies suppress inflammatory response in mice. Immunology. 2014;**143**(3):354-362. Available from: http://www.ncbi.nlm.nih.gov/pubmed/24801735

[112] Karmakar S, Bhaumik SK, Paul J, De T. TLR4 and NKT cell synergy in immunotherapy against visceral leishmaniasis. PLoS Pathogens. 2012;**8**(4):e1002646. Available from: http://dx.plos.org/10.1371/journal.ppat.1002646

[113] Barboza R, Lima FA, Reis AS, Murillo OJ, Peixoto EPM, Bandeira CL, et al. TLR4-mediated placental pathology and pregnancy outcome in experimental malaria. Scientific Reports. 2017;**7**(1):8623. Available from: http://www.ncbi.nlm.nih.gov/pubmed/28819109

[114] Mitchell RA, Altszuler R, Frevert U, Nardin EH. Skin scarification with *Plasmodium falciparum* peptide vaccine using synthetic TLR agonists as adjuvants elicits malaria sporozoite neutralizing immunity. Scientific Reports. 2016;**6**(1):32575. Available from: http://www.nature.com/articles/srep32575

[115] Abhyankar MM, Noor Z, Tomai MA, Elvecrog J, Fox CB, Petri WA. Nanoformulation of synergistic TLR ligands to enhance vaccination against *Entamoeba histolytica*. Vaccine. 2017;**35**(6):916-922. Available from: https://linkinghub.elsevier.com/retrieve/pii/S0264410X16312907

[116] Zeytun A, Chaudhary A, Pardington P, Cary R, Gupta G. Induction of cytokines and chemokines by toll-like receptor signaling: Strategies for control of inflammation. Critical Reviews in Immunology. 2010;**30**(1):53-67. Available from: http://www.ncbi.nlm.nih.gov/pubmed/20370620

[117] Zhao Y, Zhang S, Jiang L, Jiang J, Liu H. Preventive effects of *Schistosoma japonicum* ova on trinitrobenzenesulfonic acid-induced colitis and bacterial translocation in mice. Journal of Gastroenterology and Hepatology. 2009;**24**(11):1775-1780. Available from: http://doi.wiley.com/10.1111/j.1440-1746.2009.05986.x

[118] Zhang R, Sun Q, Chen Y, Sun X, Gu Y, Zhao Z, et al. Ts-Hsp70 induces protective immunity against *Trichinella spiralis* infection in mouse by activating dendritic cells through TLR2 and TLR4. PLoS Neglected Tropical Diseases. 2018;**12**(5):e0006502. Available from: https://dx.plos.org/10.1371/journal.pntd.0006502

Chapter 4

TLR Signaling on Protozoan and Helminthic Parasite Infection

Chandrani Fouzder, Alpana Mukhuty, Snehasis Das and Dipanjan Chattopadhyay

Abstract

Toll-like receptors (TLRs), a major component of innate immune system, are expressed as membrane or cytosolic receptors on neutrophils, monocytes, macrophages, dendritic cells (DCs), B lymphocytes, Th1, Th2, and regulatory T lymphocytes. It recognizes pathogen-associated molecular patterns (PAMPs) and Toll-interleukin1 (IL-1) receptor (TIR) of various invading pathogens. Downstream signaling of TLRs activates NF-κB, which acts as a transcription factor of pro-inflammatory cytokines, chemokines, and costimulatory molecules. A balance between pro- and anti-inflammatory cytokine protects host body from infectious agents and also induces the healing process. Some of parasitic infections by protozoans and helminths such as Malaria, Leishmaniasis, Trypanosomiasis, Toxoplasmosis, Amoebiasis, Filariasis, Schistosomiasis, Ascariasis, Taeniasis, and Fasciolosis are the leading cause of death and economic loss in both developing and developed nations. Frequent exposure to parasites, immigration, refugee resettlement, increasing immunodeficiency, climate change, drug resistance, lack of vaccination, etc. are the major cause of emerging and re-emerging of the above-stated diseases. However, TLR activation by parasites could stimulate antigen presenting cells and ultimately clear the pathogens by phagocytosis. So, a better understanding of host-parasite interaction in relation to TLR signaling pathway will improve the controlling method of these pathogens in immunotherapy.

Keywords: Toll-like receptors, pathogen-associated molecular patterns, protozoan parasite, helminth infection

1. Introduction

Increasing cases of parasitic infections (due to protozoans and helminths) and high rate of mortality are the greatest problem of today's world. Some of these diseases such as Malaria, Filariasis, Trypanosomiasis, Leishmaniasis, Toxoplasmosis, Amoebiasis, Ascariasis, Schistosomiasis, and Taeniasis affect over half a billion people worldwide and cause economic loss in both developing and developed countries [1]. Overpopulations, migration of people into large urban areas, and unhygienic environment are the main reasons for making these diseases epidemic [2]. However, the tragedy is that only 5% of total health expenditure was given for research work on parasitic diseases [3]. Currently, there is no effective vaccine available for these major problems. So, a better understanding of pathogenesis during infection, resistance mechanism of pathogens, host protective immune response initiation, and progression is needed for developing effective vaccines or therapeutic interventions [4].

Among the two types of vertebrate immune system, innate immunity provides the first line of defense against parasites. Previous studies stated innate immunity as nonspecific response, and it induces the acquired immunity (slower and specific response) by providing pathogens to T and B cells [5]. However, recent evidence proved that innate immune system also had a great degree of specificity and can provide host defense against invading parasites. This is because of the presence of five classes of pattern recognition receptors: TLRs (Toll-like receptors), C-type lectin receptors, NOD-like receptors (nucleotide-binding oligomerization domain leucine-rich repeat-containing receptors), RIG-I (retinoic acid inducible gene I protein) helicase receptors, and cytosolic dsDNA sensors [6, 7]. Among them, TLRs form a bridge between innate and adaptive immunity and play a very important role in parasite eradication. TLRs recognize specific pathogen-associated molecular patterns (PAMPs) in pathogens and initiate opsonization, phagocytosis, pro-inflammatory and anti-inflammatory response, and apoptosis [7, 8].

2. Cells expressing TLRs

TLRs, a major component of innate immunity, are Type-1 transmembrane glycoproteins present in both vertebrates and invertebrates [9]. Toll-like receptors are named due to their similarity with Drosophila Toll protein (Toll) [10]. All TLRs have a highly variable extracellular domain containing leucine-rich repeat (LRR) domain for ligand binding and intracellular TIR homology domain [11]. Toll-like receptors and interleukin-1 receptor together form "Interleukin-1 receptor/Toll-like receptor" superfamily whose all members have a common Toll-IL-1 receptor (TIR) domain [12]. Till date, 10 humans and 12 mice functional TLRs were identified. Although humans and mice have similar TLR1–9, TLR10 is nonfunctional in mice and TLR11–13 are lost in humans [13]. TLR1, TLR2, TLR4, TLR5, and TLR6 recognize extracellular PAMPs, which are expressed on cell surface, whereas TLR3, TLR7, TLR8, and TLR9 are expressed within endoplasmic reticulum (ER), endosomes, lysosomes, and endolysosomes and identify nucleic acids [14]. The presence of TLRs on specific intracellular vesicles restricts their activation by self-nucleic acids released by apoptotic cells [15]. TLR11 (a relative of TLR5) and TLR13 are expressed in intracellular vesicles [16], but cognate PAMP of TLR13 has not been identified yet [17]. **Table 1** shows the distribution of various TLRs in different cells.

TLRs can be classified on the basis of their recognized ligands—TLR1/TLR2 heterodimer (triacylated lipopeptides), TLR2/TLR6 heterodimer (diacylated lipopeptides), TLR4 (lipopolysaccharide), TLR3 (double-stranded RNA), TLR5

Cells	Expressing TLRs
Neutrophils	TLR 1, 2, 4, 5, 6, 7, 8
Monocytes/macrophages	TLR 1, 2, 4, 5, 6, 7, 8
Myeloid dendritic cells	TLR 2, 3, 4, 7, 8
Plasmacytoid dendritic cells (PDCs)	TLR 1, 6, 7, 9
B lymphocytes	TLR 1, 3, 6, 7, 9, 10
T lymphocytes (Th1/Th2)	TLR 2, 3, 5, 9
T lymphocytes (regulatory)	TLR 2, 5, 8
Peripheral blood mononuclear cell (PBMC)	TLR 2, 4, 5, 7, 8, 9

Table 1.
Different Toll-like receptors expressed by immune cells [7, 16].

(flagellin), TLR 7/8 (single-stranded RNA), and TLR9 (unmethylated CpG motif) [18, 19]. These ligands for TLRs are of bacterial, viral, protozoan, fungal, and helminth membrane bound or endogenously released molecules such as hyaluronic acid, fibrinogen, fibronectin, b-defensins, heparan sulfate proteoglycans, heat shock proteins, nucleic acids, and synthetically derived molecules [20].

3. TLR signaling pathway

TLRs present on dendritic cells (DCs) [both myeloid DCs (mDCs) and plasmacytoid DCs (pDCs)], neutrophils, macrophages, natural killer (NK), and natural killer T (NKT) cells induce dendritic cell maturation, MHC molecule upregulation, and costimulatory molecule production (CD40, CD80, and CD86) [21, 22]. The cytokines released by TLR signaling ultimately activate Th1 cells (via IL-12 from DCs) and Th2 cells (via IL-4 from B cell) [21, 23].

Toll-interleukin-1 receptor (TIR) domain is responsible for transducing the signal from TLRs to their adaptor proteins. The C-terminus of all TLRs, IL-1 and IL-18 and adaptor proteins of TLRs have this TIR domain. Six adaptor proteins involved in TLR signaling are MyD88 (myeloid differentiation factor 88), TIRAP (Toll-IL-1 receptor domain-containing adaptor protein) and MAL (MyD88 adapter-like), TRIF (TIR domain-containing adaptor inducing interferon-β) and TICAM-1, TRAM (TRIF-related adaptor protein) and TICAM-2, SARM (sterile-α and HEAT/Armadillo motifs-containing protein) and MyD88-5, and BCAP (B Cell Adaptor for PI3K) [24]. TLR signaling occurs via two separate pathways: MyD88 (myeloid differentiation primary response protein)-dependent pathway and MyD88-independent pathway. MyD88-dependent pathway stimulates all TLRs except TLR-3, which gets stimulated by MyD88-independent pathway. However, in case of TLR4, both MyD88-dependent and independent pathways operate [25]. MyD88 (an adaptor molecule) activates IRAK-4 (interleukin-1 receptor-associated kinase-4) alone or in combination with TIRAP (Toll-IL-1 receptor domain-containing adaptor protein) or MAL (MyD88

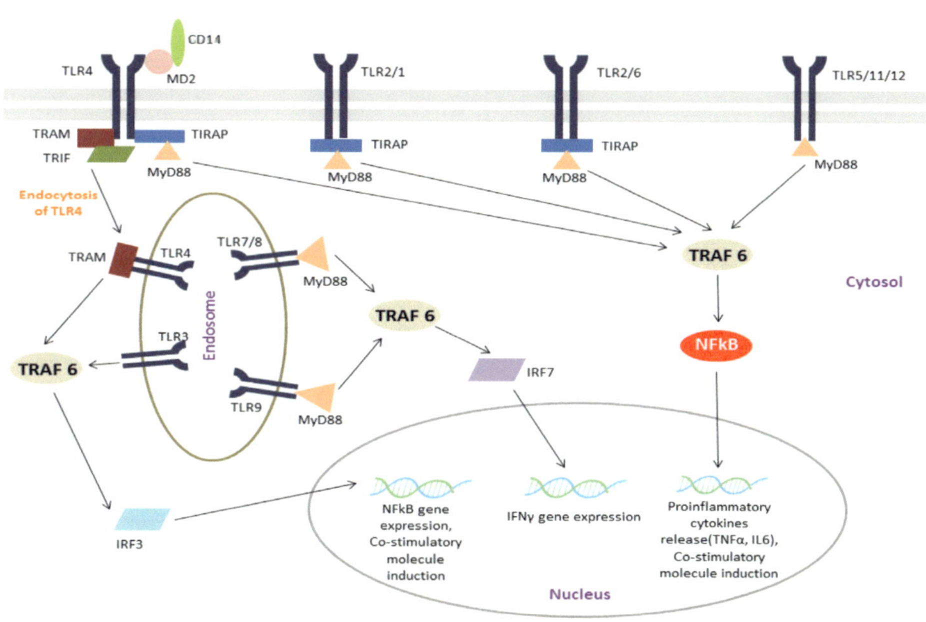

Figure 1.
TLR signaling pathway.

adapter-like). Then, IRAK-4 phosphorylates IRAK-1 [26] which in turn phosphorylates IRAK-2. IRAK-2 ubiquitinates TRAF6 (tumor necrosis factor receptor-associated factor 6) and induces two signaling pathways: (1) AP-1 (activator protein 1) activation via MAK 4/7 (mitogen-activated protein kinase) phosphorylation and (2) TAK1 (transforming growth factor-β-activated kinase 1) activation ultimately leads to MAPK (mitogen-activated protein kinase) and IKK complex [27] stimulation and nuclear factor κB (NF-κB) translocation inside the nucleus via degradation of its inhibitor. Both AP-1 and NF-κB induce the expression of pro-inflammatory cytokines and chemokines. A different MyD88-dependent pathway stimulates TLR 7, 8, and 9, which acts as a ligand for viral nucleic acids. MyD88-associated IRAK1 (interleukin-1 receptor-associated kinase-1) phosphorylates IRF7 (interferon-regulatory factor-7), which regulates Type I interferon expression [28]. TLR signaling through MyD88-independent pathway occurs via two adaptor molecules—TRIF (Toll-IL-1 receptor domain-containing adaptor inducing interferon-β) and TRAM (TRIF-related adaptor molecules) (**Figure 1**). This induces Type 1 interferon by IRF-3 (interferon-regulatory factor-3), NF-κB activation, and expression of co-stimulatory molecules [29].

4. Protozoan infections

Different protozoan (Plasmodium, Leishmania, Trypanosoma, Toxoplasma, and Entamoeba) PAMPs induced pathogenic reactions through TLR signaling pathway.

4.1 Malaria

Malaria, one of the most life-threatening diseases of human history, has infected about 219 million people over 90 countries with around 1 million deaths per year. Plasmodium, an intracellular protozoan parasite, is the causative agent of malaria. It is transmitted by infected female Anopheles mosquito biting, and four species of *Plasmodium* are responsible for human malarial infection. Among *Plasmodium falciparum*, *Plasmodium vivax*, *Plasmodium ovale,* and *Plasmodium malariae*, *P. falciparum* is the deadliest. Recently, another species named *Plasmodium knowlesi* has been found to infect humans [30]. In the early presymptomatic stage, a very low level of plasmodium can induce inflammatory response [31]. The innate immune genes such as TLRs, PRRs, and inflammatory cytokines are already upregulated, and these lead to elevate the level of TNF, IFN, and IL-12 from plasmodium-infected peripheral blood mononuclear cells (PBMCs) up to 48 h of infection [32, 33]. These inflammatory responses are associated with the pathophysiological condition and clinical symptoms of malaria including anemia, cerebral malaria, and ultimate death [34]. Cerebral malaria is caused due to overexpression and binding of adhesion molecules such as intracellular adhesion molecule 1 (ICAM-1), vascular cellular adhesion molecule 1 (VCAM-1), endothelial/leukocyte adhesion molecule (ELAM-1), and CD36 [35] on brain endothelial cell receptors. Thus, inflammatory response leads to sequestration of infected red blood cells in host brain [36]. Furthermore, TNF and IFN suppress hematopoiesis and lead to anemia during malarial infection [37]. The potential immunomodulators of the malarial parasites are: (1) plasmodial glycosylphosphatidylinositol (GPI) anchors, (2) hemozoin, and (3) plasmodial DNA. All of these three molecules are referred as "malaria toxin" released during schizogony and cause inflammation and symptoms of malaria [38, 39].

Homodimer of TLR4 and heterodimer of TLR1/TLR2 and TLR4/TLR6 can bind to GPIs released during erythrocytic phase of *P. falciparum* infection [40]. GPI induces TLR-mediated proinflammatory cytokines (TNFα and IL-1) [41] and nitric oxide [42] release from macrophages. It also induces cerebral malaria at later course of infection [43]. Plasmodium 2-Cys peroxiredoxin also acts as a TLR4 ligand in monocyte and mast cells and causes cytokine production [44]. Hemozoin is released during each life cycle of *P. falciparum* infection and makes a complex with plasmodial DNA. This complex acts as a TLR9 ligand and leads to the production of proinflammatory cytokines (TNFα and IL-1β) [45]. Hemozoin DNA complex induces cerebral malaria by caspase 1-mediated inflammasome (NLRP3) formation by TLR9 in case of *P. chabaudi* infection [46] but is absent in *P. berghei* spoprozoite infection [47]. Although in case of both mice and humans, Plasmodium infection renders no TLR stimulation in dendritic cells. The infant exposed to TLR-mediated cytokine profiles (IL-10) is associated with higher risk of *P. falciparum* maternal infection during delivery [48]. RNA of *P. chabaudi* acts as a ligand for TLR7 and induces IFNγ, IL-10, IL-12, and TNF release at 24 h of infection [49]. In case of *P. vivax* infection, TLR5 and TLR7 hinder parasitic growth, but TLR9 is associated with high inflammation and cytokine production [50]. The 19 kDa C-terminal fragment of merozoite surface protein 1 (MSp1) in *P. vivax* acts as a ligand for TLR5 [51]. *P. yoelii* infection in peritoneal macrophages enhances TLR and parasite-specific immune response [52] (**Figure 2**). Other than Th1 response, malaria parasite-derived molecules also induce Th2 response via IL-4-inducing factor (released by PI3K-Akt-NF-κB signaling) in DC [53].

4.2 Leishmaniasis

Leishmaniasis is one of the deadliest parasitic infections with an estimation of 200,000–400,000 worldwide infections each year. A protozoan parasite is the causative agent of this disease, which is transmitted to humans by the biting of female Phlebotomus sandfly. The pathology of this infection and causative parasitic species includes cutaneous (i.e., *L. major*, *L. mexicana*, and *L. guyanensis*), mucocutaneous (i.e., *L. amazonensis* and *L. braziliensis*), or visceral leishmaniasis (*L. donovani* and *L. chagasi*) [54]. Several reports indicate that few Leishmania-derived molecules could interact with innate immune receptors (TLRs) of host and result in inflammatory

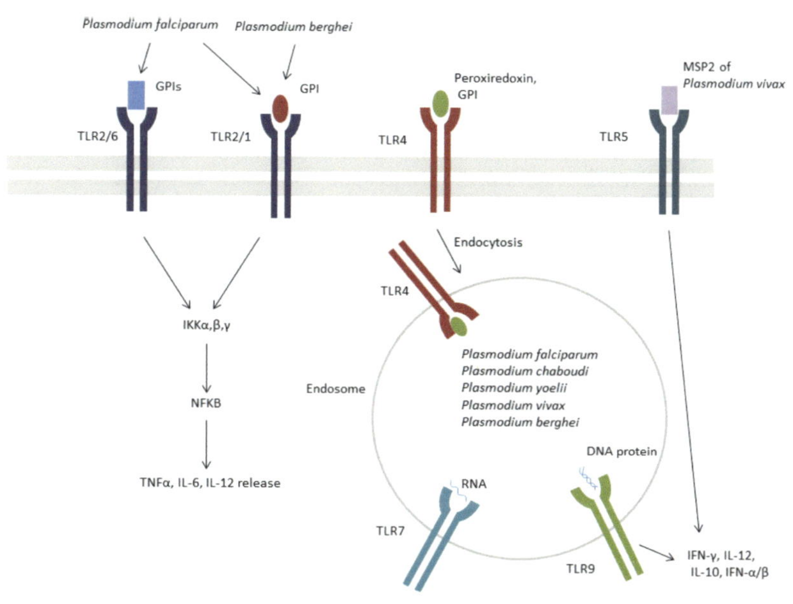

Figure 2.
TLR signaling during Plasmodium infection.

response. This inflammation effectively deprives parasites from the host by inducing efficient adaptive responses.

Lipophosphoglycan (LPG) occurs as a surface protein of *L. major*, *L. mexicana*, *L. aethiopica*, and *L. tropica* and acts as a major ligand for TLR-mediated host immune response. LPG's secreted form is structurally similar to membrane bound form with differences in sugar types of glycan and in number of phosphorylated oligosaccharide repeats. Membrane bound LPG induces ROS production and Th2 cell differentiation, whereas soluble LPG causes Th1-promoting cytokine production [55]. Inside NK cells, LPG of *L. tropica* induces TNFα, IFNγ, nitric oxide (NO), and reactive oxygen species (Th1 response) release via TLR2 upregulation and stimulation [56, 57]. TLR2 can also induce immune response by altering TLR9 expression [58]. However, in case of *L. braziliensis* and *L. amazonensis*, parasite could decrease IL-12 production, increase IL-10 production by TLR2-mediated p38 MAPK inhibition in macrophages, and thus increase pathogenesis. TLR2/TLR4 dimerization induces the expression of SOCS-1 and SOCS-3 (suppressor of cytokine signaling protein) by LPG [59]. A protein structurally related to silent information regulator 2 (SIR2) family could activate B lymphocytes, major histocompatibility complex (MHC) II, CD40 and CD86 (costimulatory molecules) overexpression, DC maturation, and TNFα and IL12 secretion through TLR2 [60]. HO-1 (heme oxygenase-1) mediated inhibition of TLR2, 4, 5, and 9 (but not TLR3) association with their adaptor proteins resulted in downregulation of TNFα and IL-12 production in *L. chagasi* and *L. donovani* infection [61]. This inflammatory imbalance occurs due to MAPKp38 phosphorylation inhibition and ERK 1/2 phosphorylation activation in macrophages. In addition, *L. donovani*, *L. mexicana* (expressed p8 proteoglycolipid complex), and *L. major* suppressed TLR4 activation by releasing TGFβ that activates A20, a complex deubiquitinating enzyme, through SRC homology region-2 domain containing phosphatase-1 (SHP-1) and IRAK inactivation [62]. Proteoglycolipid complex (P8), host-derived Apolipoprotein E (ApoE), and four glycolipids of *L. pifanoi* amastigote were the ligands of TLR4 and control the parasite [55]. P8 activates TLR4 of parasitophorous vacuole, which induces IL1 and TNFα production and aids in phagocytosis of *L. pifanoi*. At early stage of infection, neutrophil-derived serine protease and elastase results in parasite death, but at later stage, bone marrow derived macrophages (M2b macrophage) phagocytose neutrophil and helps in *L. major* replication by Th2-type response [63]. *L. panamensis* infection results in TNF α production through TLR-1, TLR-2, TLR-3, and TLR-4 pathway in human primary macrophages [64], metacyclic promastigote of *L. mexicana* induce phosphorylation of MAP kinases (ERK, p38, and JNK) through TLR4 and MΦ(bone marrow-derived macrophages), iNOS, cyclooxygenase-2 (COX-2), prostaglandin E2 (PGE2), NO, and arginase-1 are act as the inflammatory response mediators [65]. Leishmania parasites grow inside the phagolysosome of the host cells, which reflect that endosomally localized TLRs are also involved in pathogenesis [66]. In case of *L. donovani* infection, TLR7 activates IRF-5 and induces Th1 responses of host [67]. Cytosine-phosphate-guanosine motifs in DNA of *L. major* induce TLR9-mediated NK cell activation and IL12 production from bone marrow-derived DC [68, 69]. Recent reports show that viral RNA present in *L. guyanensis* (LRV1-Lg), *L. major* (LRV2-Lmj) [70], and *L. aethiopica* (LRV2-Lae) serves as a ligand for TLR3 [71]. TLR3 produces NO and TNFα during *L. donovani* infection and mediates leishmanicidal activity [72] (**Figure 3**).

4.3 Trypanosomiasis

The protozoan parasites of the genus *Trypanosoma* cause a group of disease in several vertebrates, called trypanosomiasis or trypanosomosis. In humans,

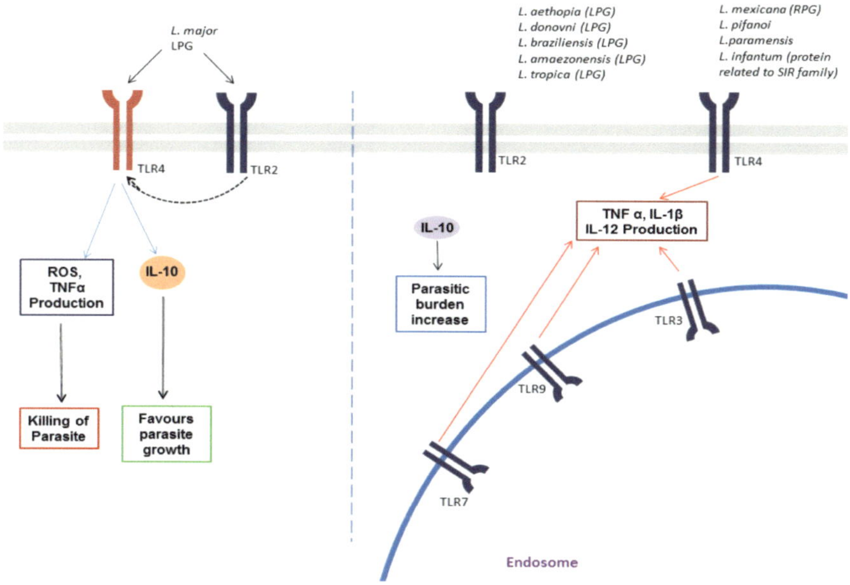

Figure 3.
TLR signaling induced by different Leishmania ligands.

Trypanosoma brucei gambiense and *Trypanosoma brucei rhodesiense* cause African trypanosomiasis or sleeping sickness (transmitted by tsetse fly), and *Trypanosoma cruzi* causes American trypanosomiasis or chagas disease (blood feeding Triatominae bugs) [73]. All of these parasites cause millions of death per year in both sub-Saharan African and Latin American countries. The disease remains asymptomatic for several years and ultimately affects the central nervous system, heart, and GI tract [74].

TLR receptor plays an important role in internalization of the parasite through phagocytosis and induces immune response for parasite eradication from cells [75]. GPI anchored mucin-like glycoproteins (tGPI-mucin contains unsaturated alkyl-acylglycerol) of the *T. cruzi* trypomastigote membrane activates MAPK (by phosphorylation) and IkB (inhibitor of NF-κB), which triggers TLR2-mediated cytokine production by macrophages [76]. A TLR2-TLR6 and CD14 complex recognize the free GPI (glycoinositophospholipids containing ceramide) from *T. cruzi* parasite (epimastigote) [77]. Tc25, a *T. cruzi* derived protein, induces TLR2-mediated pro-inflammatory cytokine release from host cells [78]. However, role of GPI anchors VSGs of *T. brucei* Trypomastigotes in specific TLR-mediated macrophage activation and proinflammatory cytokine (TNFα, IL-6, and NO) production have not been elucidated yet [79]. *T. cruzi* and *T. brucei* genomic DNA (contains unmethylated CpG motifs) have TLR9-mediated TNFα and IFNα/β stimulation and penetration of T cells in brain parenchyma [80, 81] (**Figure 4**).

4.4 Toxoplasmosis

Toxoplasma gondii, an obligate intracellular apicomplexan parasite, is a leading cause of food borne disease in a wide range of worm-blooded animals worldwide [82]. *T. gondii* causes asymptomatic toxoplasmosis in healthy adults and produces severe toxoplasmic encephalitis in immune compromised people [83]. Moreover, it causes congenital toxoplasmosis in fetus leading to death and abortion [84]. Inside its intermediate host humans, mice, etc., *T. gondii* proliferates asexually to form tachyzoite and bradyzoite stages [85].

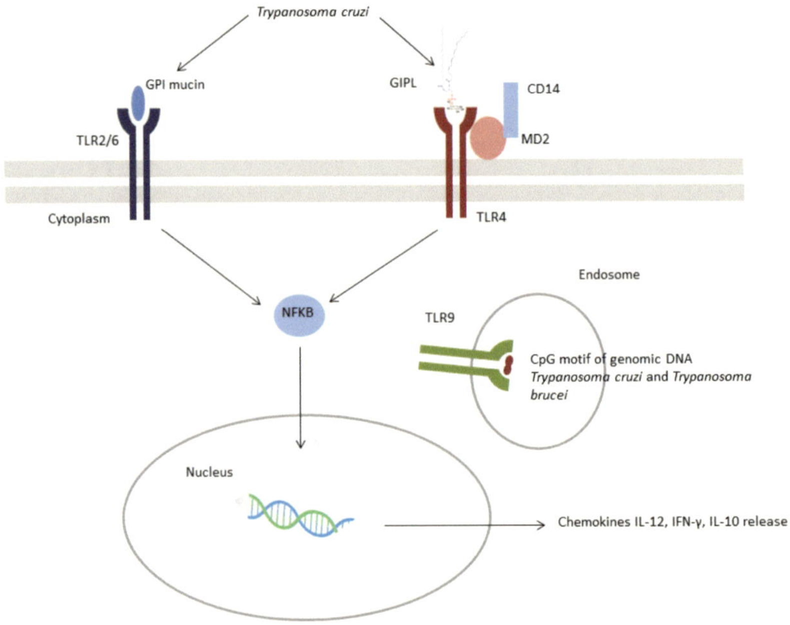

Figure 4.
Trypanosoma PAMPs and TLR signaling.

TLR11 and TLR12 recognize *T. gondii* profiling (TgPRF) and induce IL12 and IFNα production in conventional dendritic cells (cDCs), macrophages, and plasmacytoid dendritic cells (pDCs). This IFNα induces IFNγ production from NK

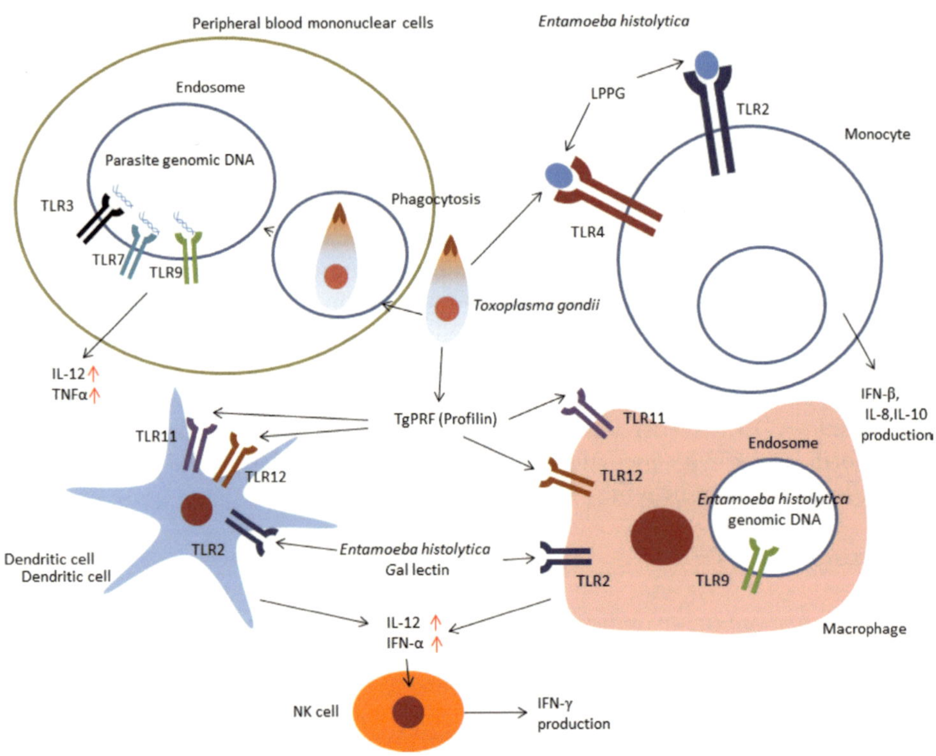

Figure 5.
Toxoplasma and Entamoeba induced TLR signaling pathway.

cells. *T. gondii* infection also induces IFNβ production in inflammatory monocytes (IMs) and TLR4-mediated phagocytic uptake of the parasite [86]. Endosomal TLRs (TLR3, TLR7, and TLR9) stimulate IL12 production in human PBMCs in response to DNA and mRNA of *T. gondii* tachyzoites when the cells were primed with IFNγ [87]. GPIs present in parasite membrane aggravate TLR2- and TLR4-mediated TNFα production in inflammatory response [88]. In some cases, tachyzoites differentiate into bradyzoites inside the central nervous system and cause neurological and behavioral abnormalities [89]. TLR2 signaling pathway makes chronic inflammation in different central nervous system cell types [85] (**Figure 5**).

4.5 Amoebiasis

Entamoeba histolytica is a protozoan parasite, which causes amebiasis in humans. It is one of the deadliest diseases after malaria and causes almost 40,000–100,000 deaths per year in underdeveloped countries [90]. The clinical symptoms include diarrhea, dysentery, pain in lower abdomen, and liver abscess, which occur due to invasion of amoeba in host lung, heart, brain, skin, genital, etc. [91]. The lipophosphopeptidoglycan (LPPG) present on the surface of *E. histolytica* induces TLR2- and TLR4-mediated NF-κB activation and cytokine (IL-8, IL-10, IL-12p40, and TNFα) release from human monocytes [92]. The Gal/GalNAc lectin (Gal-lectin), a surface molecule of *E. histolytica,* upregulates cytokines and TLR2 genes via NF-κB and MAP kinase activation in macrophages and dendritic cells [93]. TLR9 recognizes *E. histolytica* genomic DNA and helps in TNFα production in macrophages [94] (**Figure 5**).

5. Helminth infections

Although several studies were conducted on TLR signaling in response to intracellular parasites, only a few examination reflects the interaction of helminths with TLRs.

5.1 Filariasis

Lymphatic filariasis (commonly called elephantiasis), caused by three species of nematode parasites, *Wuchereria bancrofti*, *Brugia malayi*, and *B. timori*, is a major health problem in tropical countries. During initial stage, infection remains asymptomatic. Acute condition displays local inflammation of skin, lymph nodes, and lymphatic vessels, which ultimately leads to edema in chronic condition [95]. Wolbachia, an intracellular symbiotic bacterium of filarial nematode, is the major mediator of inflammatory response in case of lymphatic filariasis and onchocerciasis [2]. WSP protein in outer membrane of *Wolbachia* sp. induces TLR2- and TLR4-mediated inflammation in macrophages and DCs [96].

In case of chronic infection, filarial nematode downregulates host immune response via TLR4-mediated T cell apoptosis [97]. Live microfilariae of *B. malayi* can downregulate mRNA and protein expression of TLR1, TLR2, TLR4, and TLR9 and activate TLR2 upon antigen stimulation on B cells and monocytes [98]. In DCs, live microfilariae and microfilarial antigen (MF Ag) diminish IL-12, IFNα, and cytokine production via inhibition of NF-κB complex formation [99]. Microfilariae infective stage (L3) of *B. malayi* also shows partial inhibition of Langerhans cells (LCs) that lead to CD4+ T cell proliferation [100]. Circulating B cells (called Breg) express TLR2 and TLR4 and maintain a worm favorable condition via induction of Treg, IL-10, and filarial-specific IgG. However, Breg-mediated response causes asymptomatic infection in initial stages but leads to secondary infection by bacteria and virus in filarial patients [92]. A phosphocholine-containing glycoprotein (ES-62) of *Acanthocheilonema viteae* (rodent filarial nematode) inhibits B and T lymphocyte activation. The secretory ES-62 inhibits TLR4-mediated IL-12 and TNFα production [101] (**Figure 6**).

5.2 Schistosomiasis

Schistosomiasis is a worldwide distributed parasitic disease caused by a flatworm, Schistosoma. It accounts for 260 million infected people in tropical and sub-tropical regions (Africa, South America, the Middle East, East Asia, and the Philippines) [102]. *S. mansoni*, *S. intercalatum*, *S. haematobium*, *S. japonicum,* and *S. mekongi* are the five species of schistosomes that cause disease in humans. *S. mansoni*, *S. japonicum,* and *S. intercalatum* are responsible for intestinal schistosomiasis, while *S. haematobium* causes urinary schistosomiasis and is most important in terms of public health [102]. Fresh water snail of the genus *Bulinus* (*S. haematobium*), *Biomphalaria* (*S. mansoni*), and *Oncomelania* (*S. japonicum*) acts as an intermediate host of Schistosoma parasites [103].

S. japonicum eggs are deposited in the liver, lung, and intestinal wall of host, which induce granulomatous inflammation and progressive fibrosis. Th cells, natural killer (NK) cells, NKT cells, myeloid-derived suppressor cells (MDSCs), and macrophages are mainly involved in fight against *S. japonicum* and its eggs [104]. Expressions of TLR1, TLR3, TLR7, TLR8, and NF-κB are greatly repressed at the initial stage of schistosomiasis. TLR3 modulates Th2 response in lung in *S. mansoni* infection and in NK cells during *S. japonica* infection [105]. *S. mansoni* is known to attenuate Th1 responses (decrease IFNγ, TNFα, IL-12, and NO) but to promote Th2 immune responses (increase IL-10 and TGFβ) [106]. Although TLR4 protects the host from Schistosoma infection, TLR2 favors the parasite growth [107]. Both SEA (soluble egg antigen) and ES products of *S. mansoni* act as a strong inducer of Th2 response [108]. It induces transcription of markers CD40 and CD86 and cytokines IFNβ, TNFα, and IL-12-p40 in mouse myeloid DCs [109]. Glycans present in *S. mansoni* induce Treg by TLR2-mediated DC differentiation and IL-10 secretion [110]. Schistosoma egg product LFNP III also stimulates IL-10 production

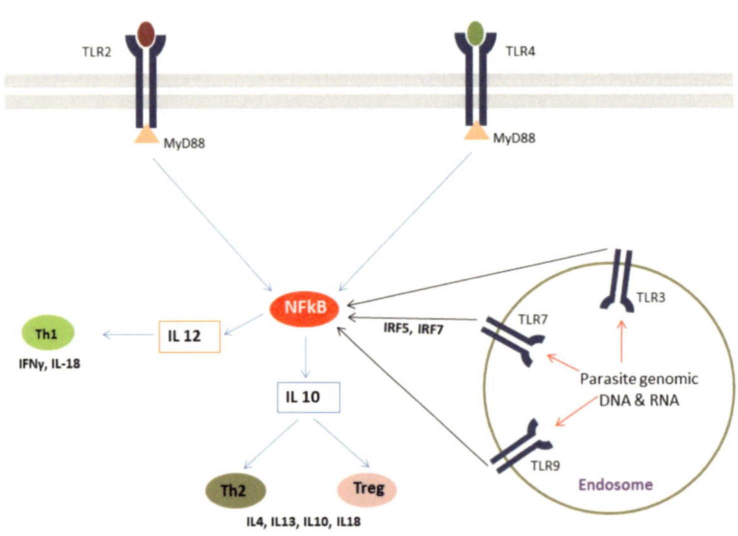

Figure 6.
TLR signaling pathway induced by Helminth pathogens.

from TLR2 and promotes Treg activation [111]. An immunomodulatory peptide, SJMHE1 of *S. japonicum*, induces TLR2-mediated Treg activation. The lysophosphatidylserine and glycolipids [112] of scistosome also activate TLR2 in DCs [113] (**Figure 6**).

5.3 Taeniasis

The pork tapeworm (*Taenia solium*) is a cestoda parasite transmitted to humans by feeding cystic larvae infected pork. Here, pig acts as an intermediate host, which swallows *T. solium* egg containing human stool and develops larva inside their body [114]. The cysticercosis cyst causes neurocysticercosis (NCC) in the nervous system, and adult taenia produces intestinal taeniasis in humans. Both are endemic in Latin America, sub-Saharan Africa, India, vast parts of China, and South East Asia [115].

TLR4 and TLR2 play an important role in developing murine NCC caused by *Mesocestoides corti* [116]. The carbohydrate of *T. crassiceps* induces TLR4- and TLR2-mediated cytokine release (IL-6 and IL-4) [117]. However, molecules derived from *T. sodium* did not induce TLR2- or TLR4-mediated cytokine release in human lymphocytes [118]. Both *T. solium* and *T. crassiceps* express several glycolipids (GSL-1) and phospholipids that may act as PAMPs. *T. crassiceps* expresses lysophosphatidylcholine [119], also present on Schistosome, and triggers TLR2 response. Although the mechanism of these molecules inducing TLR signaling has not yet been evaluated, the host may use a similar pathway of this parasite recognition [120] (**Figure 6**).

5.4 Ascariasis

Phospholipids from schistosomes and *Ascaris* worm trigger TLR2, and lysophosphatidylserine can activate DCs to induce Th2 and IL-10-producing Treg [121].

5.5 Fasciolosis

F. hepatica tegumental antigens (FhTeg), *F. hepatica* ES, and ES derived enzymes (thioredoxin peroxidase 2-Cys peroxiredoxin, fatty acid-binding protein) inhibit TLR4- and TLR3-mediated inflammatory response and facilitate parasite survival inside the host [122]. The protease activity of *F. hepatica* Cathepsin L1 (FheCL1) causes endosomal degradation of TLR3 and downregulates IL-1 production [123].

6. Conclusion

In conclusion, induction of TLR signaling pathway by infectious pathogen recognition provides a better understanding of innate immune defense mechanism against this disease. Immunotherapy emerges as a promising therapeutic approach for parasitic infection treatment over the past few years. Although no effective drugs have emerged, vaccine adjuvants yield promising results due to induction of cellular immunity via TLR. Large scales of clinical studies were conducted for developing potent and well-tolerated adjuvants. The protozoan and helminth parasites can cause activation (to a small degree) and negative regulation (to a larger degree) of TLRs resulting in increasing or decreasing parasite burden [103]. TLR agonists or antagonists are small molecule mimics, natural ligands used for treating Type I allergy, cancer, and infectious diseases. MF59 (Novartis) and AS04 (GSK) are some examples of TLR4 agonist licensed for human use [124]. GLA (TLR4 ligand) and

3M-052 (TLR7/8) ligands are now in clinical trial. Recently, RTS,S/AS01, a recombinant chimeric protein (c-terminal of circumsporozoite antigen fused with HPB antigen, and "AS01" refers to the adjuvant formulation MPL and QS21, a natural glucoside), is used for treating Malaria [125]. Several new drugs have been chemically synthesized for better understanding of the interaction of TLRs with their ligands. The knowledge from these studies will provide a greater opportunity for developing plant-derived new therapeutic drugs. So, major efforts are required for targeting TLRs in pathological conditions.

Acknowledgements

CF is thankful to the Head of the Department of Zoology, Visva-Bharati University, for extending the facilities for her research work. AM is thankful to Science & Engineering Research Board (SERB), Department of Science & Technology, Govt. of India, for her JRF fellowship (grant no. ECR/2017/001028). DC is thankful to DBT for JRF. SD thanks UGC, New Delhi, for SRF. The authors are thankful to Dr. Rakesh Kundu and Dr. Sandip Mukherjee for their technical assistance and constant encouragement.

Conflict of interest

The authors declare no conflict of interest.

Notes/thanks/other declarations

The authors thank the Head of the Department of Zoology, Visva-Bharati University, for providing the assistance in their research work.

Author details

Chandrani Fouzder[1*], Alpana Mukhuty[1], Snehasis Das[2] and Dipanjan Chattopadhyay[2]

1 Cell Signalling Laboratory, Department of Zoology, Visva-Bharati University, Santiniketan, India

2 Molecular Endocrinology Laboratory, Department of Zoology, Visva-Bharati University, Santiniketan, India

*Address all correspondence to: mum.5690@gmail.com

References

[1] Ropert C et al. Role of TLRs/MyD88 in host resistance and pathogenesis during protozoan infection: Lessons from malaria. Seminars in Immunopathology. 2008;**30**:41-51

[2] Chomel BB et al. Wildlife, exotic pets, and emerging zoonoses. Infectious Diseases. 2007;**13**:6-11

[3] Abbas AK, Lichtman AH. Cellular and Molecular Immunology; 2005;356-357

[4] Alicia K et al. Recognition of profilin by Toll-like receptor 12 is critical for host resistance to *Toxoplasma gondii*. Immunity. 2013;**38**:1-12

[5] Janeway CA Jr, Medzhitov R. Innate immune recognition. Annual Review of Immunology. 2002;**20**:197-216

[6] Akira S et al. Pathogen recognition and innate immunity. Cell. 2006;**124**:783-801

[7] Thompson MR et al. Pattern recognition receptors and the innate immune response to viral infection. Viruses. 2011;**3**(6):920-940

[8] Sivaprakasam R, et al. TLR specific immune responses against Helminth infections. Journal of Parasitology Research 2017: 9 p. Article ID: 6865789

[9] Mayank H, Veenu MH. Review. Toll-like receptors and their dual role in periodontitis: A review. Journal of Oral Science. 2011;**53**(3):263-271

[10] Medzhitov R et al. A human homologue of the Drosophila Toll protein signals activation of adaptive immunity. Nature. 1997;**388**:394-397

[11] Akira S, Takeda K. Toll-like receptor signalling. Nature Reviews. Immunology. 2004;**4**(7):499-511

[12] Dunne A, O'Neill LAJ. The interleukin-1 receptor/toll-like receptor superfamily: Signal transduction during inflammation and host defense. Science's STKE. 2003;**171**:3

[13] Kawai T, Akira S. The role of pattern-recognition receptors in innate immunity: Update on Toll-like receptors. Nature Immunology. 2010;**11**:373-384

[14] Yu L, Chen S. Toll-like receptors expressed in tumor cells: Targets for therapy. Cancer Immunology Immunotherapy. 2008;**57**:1271-1278

[15] Takeda K, Akira S. Toll-like receptors in innate immunity. International Immunology. 2005;**17**:1-1

[16] Pifer R et al. UNC93B1 is essential for TLR11 activation and IL-12-dependent host resistance to *Toxoplasma gondii*. The Journal of Biological Chemistry. 2011;**286**:3307-3314

[17] Blasius AL, Beutler B. Intracellular toll-like receptors. Immunity. 2010;**32**:305-315

[18] Doyle SL, O'Neill LA. Toll-like receptors: From the discovery of NF kappa B to new insights into transcriptional regulations in innate immunity. Biochemical Pharmacology. 2006;**72**:1102-1113

[19] Barton GM, Kagan JC. A cell biological view of Toll-like receptor function: Regulation through compartmentalization. Nature Reviews. Immunology. 2009;**9**:535-542

[20] Ulevitch RJ. Therapeutics targeting the innate immune system. Nature Reviews. Immunology. 2004;**4**:512-520

[21] Kanzler H et al. Therapeutic targeting of innate immunity with Toll-like receptor agonists and antagonists. Nature Medicine. 2007;**13**(5):552-559

[22] Ismaili J et al. Monophosphoryl lipid A activates both human dendritic cells and T cells. Journal of Immunology. 2002;**168**(2):926-932

[23] Barr TA et al. TLR-mediated stimulation of APC: Distinct cytokine responses of B cells and dendritic cells. European Journal of Immunology. 2007;**37**(11):3040-3053

[24] O'Neill LA. When signaling pathways collide: Positive and negative regulation of toll-like receptor signal transduction. Immunity. 2008;**29**:12-20

[25] Lye E et al. The role of interleukin 1 receptor-associated kinase 4 (IRAK-4) kinase activity in IRAK-4-mediated signaling. The Journal of Biological Chemistry. 2004;**270**:40653-40658

[26] Brown J et al. TLR-signaling networks: An integration of adaptor molecules, kinases, and cross-talk. Journal of Dental Research. 2011;**90**:417-427

[27] Botos I et al. The structural biology of Toll-like receptors. Structure. 2011;**19**:447-459

[28] Singh B et al. A large focus of naturally acquired *Plasmodium knowlesi* infections in human beings. Lancet. 2004;**363**:1017-1024

[29] Ockenhouse CF et al. Common and divergent immune response signaling pathways discovered in peripheral blood mononuclear cell gene expression patterns in presymptomatic and clinically apparent malaria. Infection and Immunity. 2006;**74**:5561-5573

[30] Scragg IG et al. Early cytokine induction by *Plasmodium falciparum* is not a classical endotoxin-like process. European Journal of Immunology. 1999;**29**:2636-2644

[31] Artavanis-Tsakonas K, Riley EM. Innate immune response to malaria: Rapid induction of IFN-gamma from human NK cells by live *Plasmodium falciparum*-infected erythrocytes. Journal of Immunology. 2002;**169**:2956-2963

[32] Walther M et al. Innate immune responses to human malaria: Heterogeneous cytokine responses to blood-stage *Plasmodium falciparum* correlate with parasitological and clinical outcomes. Journal of Immunology. 2006;**177**:5736-5745

[33] Clark IA, Rockett KA. The cytokine theory of human cerebral malaria. Parasitology Today. 1994;**10**:410-412

[34] Hisaeda H et al. Malaria: Immune evasion by parasites. The International Journal of Biochemistry & Cell Biology. 2005;**37**:700-706

[35] Gimenez F et al. Tumor necrosis factor alpha in the pathogenesis of cerebral malaria. Cellular and Molecular Life Sciences. 2003;**60**:1623-1635

[36] Thawani N et al. STAT6-mediated suppression of erythropoiesis in an experimental model of malarial anemia. Haematologica. 2009;**94**:195-204

[37] Krishnegowda G et al. Induction of proinflammatory responses in macrophages by the glycosylphosphatidylinositols of *Plasmodium falciparum*: Cell signaling receptors, glycosylphosphatidylinositol (GPI) structural requirement, and regulation of GPI activity. The Journal of Biological Chemistry. 2005;**280**:8606-8616

[38] Nebl T et al. Stimulation of innate immune responses by malarial glycosylphosphatidylinositol via pattern recognition receptors. Parasitology. 2005;**130**:S45-S62

[39] Durai P et al. Structure and dynamic behavior of Toll-like receptor 2 subfamily triggered by malarial

glycosylphosphatidylinositols of *Plasmodium falciparum*. The FEBS Journal. 2013;**280**:6196-6212

[40] Schofield L, Hackett F. Signal transduction in host cells by a glycosylphosphatidylinositol toxin of malaria parasites. The Journal of Experimental Medicine. 1993;**177**:145-153

[41] Zhu J et al. Induction of proinflammatory responses in macrophages by the glycosylphosphatidylinositols of *Plasmodium falciparum*: The requirement of extracellular signal-regulated kinase, p38, c-Jun N-terminal kinase and NF-kappaB pathways for the expression of proinflammatory cytokines and nitric oxide. The Journal of Biological Chemistry. 2005;**280**:8617-8627

[42] Schofield L et al. Glycosylphosphatidylinositol toxin of *Plasmodium* up-regulates intercellular adhesion molecule-1, vascular cell adhesion molecule-1, and E-selectin expression in vascular endothelial cells and increases leukocyte and parasite cytoadherence via tyrosine kinase-dependent signal transduction. Journal of Immunology. 1996;**156**:1886-1896

[43] Gazzinelli RT et al. Innate sensing of malaria parasites. Nature Reviews. Immunology. 2014;**14**:744-757

[44] Kalantari P et al. Dual engagement of the NLRP3 and AIM2 inflammasomes by plasmodium-derived hemozoin and DNA during malaria. Cell Reports. 2014;**6**:196-210

[45] Sharma S et al. Innate immune recognition of an AT-rich stem-loop DNA motif in the *Plasmodium falciparum* genome. Immunity. 2011;**35**:194-207

[46] Shio MT et al. Malarial hemozoin activates the NLRP3 inflammasome through Lyn and Syk kinases. PLoS Pathogens. 2009;**5**:e1000559

[47] Reimer T et al. Experimental cerebral malaria progresses independently of the Nlrp3 inflammasome. European Journal of Immunology. 2010;**40**:764-769

[48] Gbédandé K et al. Malaria modifies neonatal and early-life toll-like receptor cytokine responses. Infection and Immunity. 2013;**81**:2686-2696

[49] Baccarella A et al. Toll-like receptor 7 mediates early innate immune responses to malaria. Infection and Immunity. 2013;**81**:4431-4442

[50] Bargieri DY et al. New malaria vaccine candidates based on the *Plasmodium vivax* merozoite surface protein-1 and the TLR-5 agonist *Salmonella* Typhimurium FliC flagellin. Vaccine. 2008;**26**:6132-6142

[51] Bargieri DY et al. Immunogenic properties of are combinant fusion protein containing the C-terminal 19kDa of *Plasmodium falciparum* merozoite surface protein 1 and the innate immunity agonist FliC flagellin of *Salmonella* Typhimurium. Vaccine. 2010;**28**:2818-2826

[52] Fu Y et al. *Plasmodium yoelii* blood-stage primes macrophage-mediated innate immune response through modulation of toll-like receptor signalling. Malaria Journal. 2012;**11**:104

[53] Wu X et al. A malaria protein factor induces IL-4 production by dendritic cells via PI3K-AktNF-κB signaling independent of MyD88/TRIF and promotes Th2 response. The Journal of Biological Chemistry. 2018;**293**(27):10425-10434

[54] Travi B et al. Behavior of *Leishmania braziliensis* s.l. in golden hamsters: Evolution of the infection under different experimental

conditions. Journal of Parasitology. 1988;**74**(6):1059-1062

[55] Faria MS et al. Toll-like receptors in leishmania infections guardians or promoters? Journal of Parasitology Research. 2012;**2012**:930257

[56] Becker I et al. Leishmania lipophosphoglycan (LPG) activates NK cells through toll-like receptor-2. Molecular and Biochemical Parasitology. 2003;**130**:65-74

[57] Hawn TR et al. *Leishmania major* activates IL-1 alpha expression in macrophages through a MyD88-dependent pathway. Microbes and Infection. 2002;**4**:763-771

[58] Srivastava S et al. Leishmania expressed lipophosphoglycan interacts with Toll-like receptor (TLR)-2 to decrease TLR-9 expression and reduce anti-leishmanial responses. Clinical and Experimental Immunology. 2013;**172**:403-409

[59] Chandra D, Naik S. *Leishmania donovani* infection down-regulates TLR2-stimulated IL-12p40 and activates IL-10 in cells of macrophage/monocytic lineage by modulating MAPK pathways through a contact-dependent mechanism. Clinical and Experimental Immunology. 2008;**154**:224-234

[60] Silvestre R et al. The contribution of Toll-like receptor 2 to the innate recognition of a *Leishmania infantum* silent information regulator 2 protein. Immunology. 2009;**128**:484-499

[61] Saha S et al. *Leishmania donovani* exploits macrophage heme oxygenase-1 to neutralize oxidative burst and TLR signaling–dependent host defense. Journal of Immunology. 2019;**202**:1-14

[62] Srivastav S et al. *Leishmania donovani* exploits host deubiquitinating enzyme A20, a negative regulator of TLR signaling, to subvert host immune response. Journal of Immunology. 2012;**189**:924-934

[63] Filardy AA et al. Pro-inflammatory clearance of apoptotic neutrophils induces an IL-12 (low) IL-10 (high) regulatory phenotype in macrophages. Journal of Immunology. 2010;**185**:2044-2050

[64] de Veer MJ et al. MyD88 is essential for clearance of *Leishmania major*: Possible role for lipophosphoglycan and toll-like receptor 2 signaling. European Journal of Immunology. 2003;**33**:2822-2831

[65] Shweash M et al. *Leishmania mexicana* promastigotes inhibit macrophage IL-12 production via TLR-4 dependent COX-2, iNOS and arginase-1 expression. Molecular Immunology. 2011;**48**:1800-1808

[66] O'Neill LA. How Toll-like receptors signal: What we know and what we don't know. Current Opinion in Immunology. 2006;**18**:3-9

[67] Paun A et al. Critical role of IRF-5 in the development of T helper 1 responses to *Leishmania donovani* infection. PLoS Pathogens. 2011;7(1):e1001246

[68] Li Y et al. IL-18 gene therapy develops Th1-type immune responses in *Leishmania major*-infected BALB/c mice: Is the effect mediated by the CpG signaling TLR9? Gene Therapy. 2004;**11**:941-948

[69] Abou Fakher FH et al. TLR9-dependent activation of dendritic cells by DNA from *Leishmania major* favors Th1 cell development and the resolution of lesions. Journal of Immunology. 2009;**182**:1386-1396

[70] Ives A et al. Leishmania RNA virus controls the severity of mucocutaneous leishmaniasis. Science. 2011;**331**:775-778

[71] Zangger H et al. *Leishmania aethiopica* field isolates bearing an

endosymbiontic dsRNA virus induce pro-inflammatory cytokine response. PLoS Neglected Tropical Diseases. 2014;**8**(4):e2836

[72] Flandin JF et al. RNA interference reveals a role for TLR2 and TLR3 in the recognition of *Leishmania donovani* promastigotes by interferon-gamma-primed macrophages. European Journal of Immunology. 2006;**36**:411-420

[73] Bugue TA et al. Sleep structure: A new diagnostic tool for stage determination in sleeping sickness. Acta Tropica. 2005;**93**:107-117

[74] Jordan A. Tsetse flies as vectors of trypanosomes. Veterinary Parasitology. 1976;**2**:143-152

[75] Ramasawmy R et al. Heterozygosity for the S180L variant of MAL/TIRAP, a gene expressing an adaptor protein in the Toll-like receptor pathway, is associated with lower risk of developing chronic Chagas cardiomyopathy. The Journal of Infectious Diseases. 2009;**199**:1838-1845

[76] Almeida IC et al. Highly purified glycosylphosphatidylinositols from *Trypanosoma cruziare* potent proinflammatory agents. The EMBO Journal. 1999;**19**(7):1476-1485

[77] Oliveira AC et al. Expression of functional TLR4 confers proinflammatory responsiveness to *Trypanosoma cruzi* glycoinositolphospholipids and higher resistance to infection with *T. cruzi*. Journal of Immunology. 2004;**173**:5688-5696

[78] Ouaissi A et al. The *Trypanosoma cruzi* Tc52-released protein induces human dendritic cell maturation, signals via Toll-like receptor 2, and confers protection against lethal infection. Journal of Immunology. 2002;**168**:6366-6374

[79] Daniel NA et al. Distinct toll-like receptor signals regulate cerebral parasite load and interferona/band tumor necrosis factor–dependent T-cell infiltration in the brains of *Trypanosoma brucei*–infected mice. Journal of Infectious Diseases. 2012;**205**:320-332

[80] Bafica A et al. TLR9 and TLR2 signaling together account for MyD88-dependent control of parasitemia in *Trypanosoma cruzi* infection. Journal of Immunology. 2006;**177**:3515-3519

[81] Drennan MB et al. The induction of a type 1 immune response following a *Trypanosoma brucei* infection is MyD88 dependent. Journal of Immunology. 2005;**175**:2501-2509

[82] Pappas G et al. Toxoplasmosis snapshots: Global status of *Toxoplasma gondii* seroprevalence and implications for pregnancy and congenital toxoplasmosis. International Journal for Parasitology. 2009;**39**:1385-1394

[83] Luft BJ et al. Outbreak of central-nervous-system toxoplasmosis in western Europe and North America. Lancet (London England). 1983;**1**:781-784

[84] Jones JL et al. *Toxoplasma gondii* infection in the United States, 1999 2004, decline from the prior decade. The American Journal of Tropical Medicine and Hygiene. 2007;**77**:405-410

[85] Kousuke U et al. Transcriptional profiling of Toll-like receptor 2–deficient primary murine brain cell sduring *Toxoplasma gondii* infection. PLoS One. 2017;**12**(11):e0187703

[86] HA S-J et al. Internalization and TLR-dependent type I interferon production by inflammatory monocytes to *Toxoplasma gondii*. Immunology and Cell Biology. 2014;**92**(10):872-881

[87] Alan S et al. Innate recognition of *Toxoplasma gondii* in humans involves a mechanism distinct from that utilized

by rodents. Cellular & Molecular Immunology. 2017;**14**:36-42

[88] Barbalat R et al. Toll-like receptor 2 on inflammatory monocytes induces type I interferon in response to viral but not bacterial ligands. Nature Immunology. 2009;**10**:1200-1207

[89] Skariah S et al. *Toxoplasma gondii*: Determinants of tachyzoite to bradyzoite conversion. Parasitology Research. 2010;**107**:253-260

[90] Marie C, Petri WA Jr. Regulation of virulence of *Entamoeba histolytica*. Annual Review of Microbiology. 2014;**68**:493-520

[91] Maldonado-Bernal C et al. The innate immune response to *Entamoeba histolytica* lipopeptidophosphoglycan is mediated by toll-like receptors 2 and 4. Parasite Immunology. 2005;**27**(4):127-137

[92] Maldonado C, et al. Lipophosphopeptidoglycan of *Entamoeba histolytica* induces an antiinflammatory innate immune response and downregulation of toll-like receptor 2 (TLR-2) gene expression in human monocytes. Archives of Medical Research 2000;**31**(4, suppl. 4):S71-S73

[93] Kammanadiminti J et al. Regulation of Toll-like receptor-2 expression by the gallectin of *Entamoeba histolytica*. The FASEB Journal Expresse. 2003;**18**(1):155-157

[94] Catherine PA et al. Toll-like receptor 9-dependent macrophage activation by *Entamoeba histolytica* DNA. Infection and Immunity. 2008;**76**(1):289-297

[95] Brattig NW et al. The major surface protein of *Wolbachia endosymbionts* in filarial nematodes elicits immune responses through TLR2 and TLR4. Journal of Immunology. 2004;**173**(1):437-445

[96] Taylor MJ et al. Inflammatory responses induced by the filarial nematode *Brugia malayi* are mediated by lipopolysaccharide-like activity from endosymbiotic *Wolbachia* bacteria. The Journal of Experimental Medicine. 2000;**191**(8):1429-1436

[97] Martin I et al. *Fasciola hepatica* fatty acid binding protein inhibits TLR4 activation and suppresses the inflammatory cytokines induced by lipopolysaccharide in vitro and in vivo. Journal of Immunology. 2015;**194**(8):3924-3936

[98] Babu S, Nutman TB. Immunopathogenesis of lymphatic filarial disease. Seminars in Immunopathology. 2012;**34**: 847-861

[99] Venugopal PG et al. Activation and regulation of toll-like receptors (TLRs) by helminth parasites. Immunologic Research. 2009;**43**:252-263

[100] Kagan JC, Medzhitov R. Phosphoinositide-mediated adaptor recruitment controls Toll-like receptor signaling. Cell. 2006;**125**:943-955

[101] Goodridge HS et al. Immunomodulation via novel use of TLR4 by the filarial nematode phosphorylcholine-containing secreted product, ES-62. Journal of Immunology. 2005;**174**:284-293

[102] McManus DP, Loukas A. Current status of vaccines for schistosomiasis. Clinical Microbiology Reviews. 2008;**21**:225-242

[103] Sturrock RF. Schistosomiasis epidemiology and control: How did we get here and where should we go? Memórias do Instituto Oswaldo Cruz. 2001;**96**(Suppl):17-27

[104] Cha H et al. Differential pulmonic NK and NKT cell responses in *Schistosoma japonicum*-infected

mice. Parasitology Research. 2017;**116**(2):559-567

[105] Joshi AD et al. TLR3 modulates immunopathology during a *Schistosoma mansoni* egg-driven Th2 response in the lung. European Journal of Immunology. 2008;**38**(12):3436-3449

[106] La Flamme AC et al. Schistosomiasis decreases central nervous system inflammation and alters the progression of experimental autoimmune encephalomyelitis. Infection and Immunity. 2003;**71**(9):4996-5004

[107] Gao Y et al. Deficiency in TLR2 but not in TLR4 impairs dendritic cells derived IL-10 responses to schistosome antigens. Cellular Immunology. 2012;**272**(2):242-250

[108] Okano M et al. Induction of Th2 responses and IgE is largely due to carbohydrates functioning as adjuvants on *Schistosoma mansoni* egg antigens. Journal of Immunology. 1999;**163**(12):6712-6717

[109] Ritter M et al. *Schistosoma mansoni* triggers Dectin-2, which activates the Nlrp3 inflammasome and alters adaptive immune responses. Proceedings of the National Academy of Sciences of the United States of America. 2010;**107**(47):20459-20464

[110] Moreels TG et al. Concurrent infection with *Schistosoma mansoni* attenuates inflammation induced changes in colonic morphology, cytokine levels, and smooth muscle contractility of trinitrobenzene sulphonic acid-induced colitis in rats. Gut. 2004;**53**(1):99-107

[111] Velupillai P et al. B-1 cell (CD5+B220+) outgrowth in murine schistosomiasis is genetically restricted and is largely due to activation by polylactosamine sugars. Journal of Immunology. 1997;**158**(1):338-344

[112] Goodridge HS et al. Phosphorylcholine mimics the effects of ES-62 on macrophages and dendritic cells. Parasite Immunology. 2007;**29**(3):127-137

[113] Garcia HH, Del Brutto OH. Neurocysticercosis: Updated concepts about an old disease. Lancet Neurology. 2005;**4**(10):653-661

[114] Schantz PM. *Taenia solium* cysticercosis: An overview of global distribution and transmission. In: Singh G, Prabhakar S, editors. *Taenia solium* Cysticercosis from Basic to Clinical Science. Oxon, UK: CABI Publishing; 2002

[115] Garcia HH et al. Hyperendemic human and porcine *Taenia solium* infection in Peru. The American Journal of Tropical Medicine and Hygiene. 2003;**68**(3):268-275

[116] Mishra BB et al. MyD88-deficient mice exhibit decreased parasite-induced immune responses but reduced disease severity in a murine model of neurocysticercosis. Infection and Immunity. 2009;**77**(12):5369-5379

[117] Dissanayake S, Shahin A. Induction of interferon-gamma by *Taenia crassiceps* glycans and Lewis sugars in naive BALB/c spleen and peritoneal exudate cells. Molecular Immunology. 2007;**44**(7):1623-1630

[118] Uddin J et al. Mechanisms regulating monocyte CXCL8 secretion in neurocysticercosis and the effect of antiparasitic therapy. Journal of Immunology. 2010;**185**(7):4478-4484

[119] Baumeister S et al. Comparative serological reactivity of *Taenia crassiceps*, *Taenia solium* and *Taenia saginata* metacestode neutral glycolipids to infection serum from *Taenia crassiceps*-infected mice. Molecular and Biochemical Parasitology. 1992;**53**(1-2):53-61

[120] Mills GL et al. Lipid composition of metacestodes of *Taenia taeniaeformis* and lipid changes during growth. Molecular and Biochemical Parasitology. 1981;**3**(5):301-318

[121] Bradley JE, Jackson JA. Trichuriasis and ascariasis review article immunity, immunoregulation and the ecology of trichuriasis and ascariasis. Parasite Immunology. 2004;**26**:429-441

[122] Motran CC et al. Helminth infections: Recognition and modulation of the immune response by innate immune cells. Frontiers in Immunology. 2018;**9**:664(1-12)

[123] Donnelly S et al. Helminth cysteine proteases inhibit TRIF-dependent activation of macrophages via degradation of TLR3. The Journal of Biological Chemistry. 2010;**285**:3383-3392

[124] Fox CB et al. A nanoliposome delivery system to synergistically trigger TLR4 AND TLR7. Journal of Nanbiotechnology. 2014;**26**(12):17

[125] Kalantari P. The emerging role of pattern recognition receptors in the pathogenesis of malaria. Vaccine. 2018;**6**:13

Chapter 5

Toll-Like Receptors (TLRs) in Neurodegeneration: Integrative Approach to TLR Cascades in Alzheimer's and Parkinson's Diseases

Juan M. Zolezzi, Sussy Bastías-Candia and Nibaldo C. Inestrosa

Abstract

Sterile inflammatory response constitutes a main event in several neurodegenerative disorders. Alzheimer's disease (AD) and Parkinson's disease (PD), the leading degenerative pathologies of the central nervous system worldwide, exhibit a strong inflammatory component. Microglial and astrocytic reactivity, increased levels of inflammatory mediators, neuronal damage, and death are part of the pathological scenario leading to the progressive failure of the brain neuronal network. In this regard, the link between the toll-like receptors (TLRs)-mediated inflammatory cascade and the molecular hallmarks of AD and PD have been demonstrated elsewhere. Moreover, the long-lasting exposure to the inflammatory environment is considered one of the key elements leading to the establishment and progression of these pathologies. Accordingly, the modulation of the inflammatory response has emerged as a main target of new therapeutic approaches to fight these diseases. In this regard, and based on our previous works on this subject, we describe the pathological profile of both pathologies but in the inflammatory context. Thus, in the present chapter, we will introduce the main aspects of both diseases and how they interplay with the TLR-mediated response. We believe that this chapter should provide a concise overview of the roles of TLRs in the inflammatory cascades triggered during AD and PD pathophysiology.

Keywords: neurodegeneration, neuroinflammatory response, toll-like receptors, Alzheimer's disease, Parkinson's disease

1. Introduction

The aging of the world population has been demonstrated systematically by several demographic studies. Regrettably, increased life expectancy has led to an increased prevalence of age-related disorders including neurodegenerative diseases. In this regard, Alzheimer's and Parkinson's diseases constitute the most relevant issues for the public health system of different countries. Accordingly, during the last decades, significant efforts have been committed to improve our understanding of the molecular cascades responsible for an altered aging process as well as for the establishment and progression of neurodegenerative disorders, mainly Alzheimer's disease (AD) and Parkinson's disease (PD) [1].

Relevantly, though AD and PD possess their very own pathological characteristics, the neuroinflammatory milieu has emerged as a central event of the chronic degenerative process. From the prodromal stage of these disorders up to the more advanced ones, inflammation seems to accompany or, in some cases, to drive the progression of AD and PD. Importantly, inflammation, as part of the nonspecific immune response, plays a critical role in maintenance of system homeostasis, allowing to prevent or to control the detrimental effects induced by a wide variety of xenobiotics to the cellular components of the biological systems. Remarkably, from an unspecific harmful stimulus, a whole range of responses are triggered including complement cascade activation and cytokines release as well as activation of the immune cells located at the site of the insult. To properly eliminate the initial cause of distress and to repair the damaged tissue, the inflammatory response must be delicately balanced considering not only pro-inflammatory but anti-inflammatory mediators as well. Tumor necrosis factor 1 (TNF-1α), interleukins (IL-1, IL-8, IL-10), interferon (INF-γ), transforming growth factor 1 (TGF-1), complement proteins, act together to develop a coordinated response against primary, unspecific stimuli [2, 3]. In this regard, the compromise of this essential system relates with severe, and often lethal, conditions including immunodeficiency syndromes as well as autoimmune diseases. Relevantly, during the last decades, a lot of attention has been also given to the effects of chronic inflammatory condition as the starting point of different degenerative conditions. Among these, neurodegenerative disorders, such as Alzheimer's and Parkinson's diseases, have found in the inflammatory response a critical milieu of events able to determinate the main molecular and cellular events verified during each of these diseases [4].

2. Central nervous system (CNS) immunocompetence

CNS is a highly specialized structure whose functions require specific microenvironmental conditions. Moreover, neuronal activity and, thus, the health of the neuronal network depend on the maintenance of ion gradients whose concentration differs significantly from the rest of the body compartments [5]. To ensure these conditions, the CNS remains partially isolated by the existence of the blood-brain barrier (BBB), a highly complex semipermeable cellular barrier whose main function is to prevent both exogenous and endogenous elements to alter the brain homeostasis [5–7]. Regarding the immune response, microglial population constitutes the only immune system cellular representative within the brain with the astrocytes acting as a companion to exert immune surveillance and to act as the first line of response against harmful events within the brain. Although some additional peripheral immune cells including cluster of differentiation 11b and c (CD11b, CD11c)-positive cells localize to the CNS during some inflammatory conditions, it is believed that this situation is caused because of an altered permeability of the BBB as a consequence of the primary insult [8]. In this sense, the brain parenchyma has been defined as an anti-inflammatory environment due the increased levels of relevant anti-inflammatory mediators including the transforming growth factor β (TGFβ) and interleukin (IL)-10, preventing both the immune cell spreading across the CNS and an excessively strong immune response [8–10]. However, this latter condition does not imply that the CNS cannot answer to immune challenges; on the contrary, the CNS is a fully immunocompetent system, but with this mechanism being tightly controlled in order to avoid secondary damage caused by extensive inflammation and detrimental cell damage end products, such as reactive oxygen species (ROS). In this context, during the last years, it has been evidenced that sustained inflammatory challenge, whether systemic or local to the CNS, will alter significantly the neuronal environment affecting severely the neuronal function and the health of the neuronal network. Regrettably, it has been suggested that prior to AD and PD establishment as well as during their progression, a chronic inflammatory condition has developed, contributing to the molecular alterations observed during these pathologies.

3. Toll-like receptors

A significant characteristic of AD and PD is that these pathologies exhibit a strong inflammatory component even when both are sterile conditions. In this sense, the innate immune system works through the pattern recognition receptors (PRRs) which recognize molecular patterns related to pathogens (pathogen-associated molecular patterns, PAMPs) and to endogenous molecules indicative of cell damage (damage-associated molecular patterns, DAMPs), such as high-mobility group protein B1 (HMGB1), S100 proteins, heat shock proteins (HSPs), DNA, mitochondrial DNA (mt-DNA), and ATP [11–15]. The toll-like receptors (TLRs) family constitutes a highly relevant type of PRR necessary not only to unleash the initial immune response but also to connect the first unspecific defense with the secondary adaptive immunity [12].

Depending on the species, 11–13 TLR subtypes can be found. Relevantly, the localization of these receptors within the cells differs between the different TLRs. In this regard, while TLRs 1, 2, 4, 5, and 6 are expressed at the cell membrane, its main objective being to sense the extracellular compartment, the TLRs 3, 7, 8, and 9 are located inside the cells, mainly associated with endosomes and sensing the internal microenvironment for viral components, such as RNA and DNA [12, 16]. The presence of TLRs has been determined not only in several cell components of the peripheral immune system but also in the different cell types found in the brain including astrocytes, microglia, neurons, and oligodendrocytes, suggesting that each of these cell types can sense and trigger an immune response in the presence of different harmful molecular patterns. Interestingly, it has been demonstrated that not all the cells within the brain express the same pattern of TLRs. For example, microglia and neurons express all TLR subtypes, while astrocytes express a more limited repertoire, including TLR2, TLR3, TLR4, TLR9, and TLR11 [16, 17].

3.1 TLR inflammatory cascade

The signaling cascade triggered after TLRs' activation involves the cross talk with several additional pathways able to critically modify cell physiology. In this sense, canonical TLR-mediated signaling involves the myeloid differentiation factor 88 (MyD88) cascade. In this pathway, TLR activation couples with MyD88 inducing the activation of interleukin-1 receptor-associated kinase (IRAK) and, subsequently, the activation of the tumor necrosis factor receptor-associated factor 6. This event allows the recruitment of the transforming growth factor-β-activated kinase-1 (TAK1) which, together with the TAK1-binding proteins, leads to the phosphorylation of IκB causing the activation of the IKK complex and the release of the nuclear factor-κB (NF-κB), triggering the NF-κB-dependent inflammatory response [11, 18, 19]. Importantly, TLR 3 and TLR 4 can also signal via the TIR-containing adaptor inducing interferon-β (IFN-β) (TRIF). In this additional pathway, additional to the release of NF-κB, it also causes an increased expression of IFN-β by means of the IKKε/TANK-binding kinase-1 (TBK1)-dependent phosphorylation of the interferon regulatory factor 3 and 7 (IRF3 and IRF7) [7, 19, 20].

The main objective of these TLR-mediated processes is to regulate the expression of several pro-inflammatory and anti-inflammatory mediators including IL-1, IL-6, IL-10, IL-11, IL-12, tumor necrosis factor (TNF), TGF, IFN, CCL2, CCL5, CXCL8, and CXCL10, among others [4, 11, 18, 19].

Additionally, TLR activation can also signal through complementary molecular pathways. Indeed, TAK1 activation also induces nemo-like kinase (NLK) and the c-Jun N-terminal kinases (JNK) pathway [18–21]. Similarly, MyD88 can signal through the phosphatidylinositide-3 kinase (PI3K)/Akt pathway, modulating the activity of the glycogen synthase kinase 3 β (GSK3β) [22, 23]. On the other hand, it has been demonstrated that TLR2 and TLR4 can activate the PI3K/Akt pathway through the Ras-related C3 botulinum toxin substrate 1 (Rac1), a member of the Rho family of GTPases [23]. Complimentarily, the Janus kinase (JAK)/signal transducer and activator of transcription (STAT) (JAK/STAT) pathway is known to respond to a variety of PAMPs/DAMPs and cytokines, including different interleukins and INF-β. Relevantly, different studies have demonstrated that several members of the TLR family can phosphorylate different STAT members, suggesting a direct modulation of the JAK/STAT pathway [24, 25]. As is possible to observe, the molecular cascades that could be triggered secondarily to TLR activation are related with critical cellular processes including cell cycle modulation, apoptosis, and cytoskeleton remodeling, among others. This situation depicts the complexity of the immune response and the relevance of its modulation in the context of different pathologies including neurodegenerative ones, such as AD and PD.

4. Alzheimer's and Parkinson's diseases: Aβ/SNCA and TLRs

4.1 Alzheimer's disease

AD constitutes the main form of dementia in the elderly population. Although, AD is recognized by the memory impairment and the reduced cognitive performance, the clinical scenario starts with mood alterations at the very beginning of the disease followed by the compromise of the short-term memory and the loss of long-term memory as the pathology progresses. Histologically, AD is characterized by the atrophy of the frontal cortex, limbic area, and hippocampus. On the other hand, the molecular hallmarks of AD are the formation of the amyloid-β (Aβ) plaques, constituted by the aggregated forms of the amyloid-β peptide, and the intraneuronal formation of neurofibrillary tangles (NFTs) composed of hyperphosphorylated tau protein [26]. Relevantly, beyond these hallmarks, AD also exhibits increased oxidative stress, mitochondrial dysfunction, and chronic inflammatory response. Altogether, these molecular alterations will lead to synaptic damage, neuronal loss, and neuronal circuitry breakdown [26, 27].

Relevantly, even when genetic conditions can lead to an early onset AD presentation, this accounts only for a small number of cases worldwide, with over 95% of the cases being termed as sporadic or late-onset AD. In this latter case, age and lifestyle have been defined as the main risk factors associated with the appearance of the disease [26, 27]. In agreement with the amyloid hypothesis of AD, these risk factors have a direct impact on the levels of Aβ leading to its increased production and subsequent accumulation within the brain [28, 29]. In this regard, different studies have linked the synaptic failure, mitochondrial dysfunction, tau hyperphosphorylation, glial activation, and neuronal death with increased levels of Aβ [27, 30]. Importantly, in the context of neuroinflammation, the ability of Aβ to induce the inflammatory response through the TLR2, TLR4, and TLR6, as well as their co-receptors including CD36, CD14, and CD47 has been widely demonstrated

[31–33]. However, it must to be noticed that systemic inflammatory conditions have been linked with an increased risk of AD development, suggesting that the pro-inflammatory mediators are able to alter brain homeostasis leading to neuronal damage and to the beginning of the neurodegenerative process [34–38].

4.2 Parkinson's disease

PD corresponds to the second most common neurodegenerative disorder. PD is characterized by dopaminergic circuitry impairment caused by the loss of the dopaminergic neurons, mainly at the substantia nigra pars compacta (SNpc). Defined as synucleinopathy, PD exhibits neuronal inclusions of aggregated α-synuclein (SNCA) protein which can be located at the cell soma and/or neurites, forming the Lewy bodies and Lewy neurites [39]. SNCA plays relevant roles in the synaptic activity as it is linked with the recycling of synaptic vesicle pools [40–42]. Moreover, SNCA interferes with the dopamine metabolism affecting both the tyrosine hydroxylase, the enzyme in charge of synthetizing dopamine, and the dopamine transporter (DAT) [43, 44]. Similar to AD, even when the main signs of PD are associated with motor impairment, the pathological scenario begins much earlier with sleep disturbances and loss of olfaction which often passes unadvised to the patients or their relatives. Dementia is also an additional feature of the pathology and the affectation of memory can be part of the clinical picture. Although SNCA aggregation explains the damage to the dopaminergic neurons, the mechanisms behind SNCA dynamics have remained elusive [39, 45, 46]. Although PD can also emerge as a genetics-related disease with several genes linked to an early presentation, only a small proportion of the cases share this condition worldwide [39, 47]. In the same way, the vast majority of PD cases are linked to aging and lifestyle with the exposure of the patients to chemical xenobiotics, such as pesticides and recreational drugs being of most relevance [48]. Importantly, the SNCA aggregation is also related to the additional features observed during the pathological process, including mitochondrial dysfunction, increased oxidative stress, and neuroinflammation. Moreover, oxidative stress plays a central role in the pathophysiology of PD and the contribution of the ROS to the inflammatory milieu seems to be part of the establishment and progression of the disease.

Regarding the SNCA and inflammatory triggering, different researchers have demonstrated that depending on the aggregation status of SNCA, this protein can activate the TLRs, at least TLR2 and TLR4 [49–52]. Moreover, SNCA can be incorporated by the surrounding cells, particularly astrocytes, leading to the formation of SNCA inclusions also in these latter cells and helping to spread the SNCA pathology across the brain [53, 54]. Relevantly, PD can also be influenced by the systemic inflammatory status. Indeed, it has been recently demonstrated that increased levels of inflammatory mediators favor SNCA aggregation [55, 56].

4.3 Aβ/SNCA secondary inflammatory cascade

Although we have indicated that both Aβ and SNCA can activate the TLRs, we have only spoken about the direct activation induced by these molecules on some representatives of the TLR family. However, we must realize that once the TLRs are activated, a full repertory of pro-inflammatory mediators is released to the environment. In this context, Aβ induces the expression of IL-1, IL-6, IL-12, TNF-α, cyclooxygenase 2 (COX2), and the inducible nitric oxide synthase (iNOS) [57]. Additionally, because of the cellular damage caused, it will also induce the release of further DAMPs [26, 27, 58]. Similarly, TLRs activated after SNCA challenge will increase the expression of TNF-α, IL-6, and CXCL1 [49–51]. In the case of SNCA, the cell damage also will cause the release of several DAMPs. Relevantly, the pro-inflammatory mediators and the subsequent DAMPs induced by Aβ and SNCA are able to further activate additional members of the TLR family, enhancing the inflammatory response. If we take into account that in both pathologies the levels of Aβ and SNCA are steadily increasing, the concept of a chronic inflammatory condition emerges as a potential mechanism to explain the progression of both diseases. Moreover, some of these pro-inflammatory mediators can also have a direct impact on the neuronal activity. Such is the case of the glial TNF-α-mediated expression of the AMPA receptors within the postsynaptic terminal. In this case, the increased production and release of TNF-α by the astrocytes, perhaps induced by the chronic exposure to the inflammatory stimulus, will cause the hyperexcitability of the neurons leading to glutamate excitotoxicity [59–62].

4.4 Microglial priming

Relevantly, an additional effect caused by Aβ and SNCA should be considered. It has been demonstrated that both molecules are also able to induce a phenomenon termed "microglial priming." In this regard, microglial population which remains in a resting state when exposed to different inflammatory mediators, DAMPs, and/or PAMPs can differentiate into two activated phenotypes, the M1 and M2. While the M1 is considered as a pro-inflammatory activation state, the M2 is defined as the anti-inflammatory microglial phenotype. Interestingly, it has been evidenced that in the presence of INF-γ and the TLR-mediated signaling, microglia usually undergo M1 transformation. Moreover, when microglia became "primed" usually changes to the M2 phenotype but develops a significant sensibility to new exposures to harmful stimuli, exhibiting an over dimensioned response and causing the abnormal raising of pro-inflammatory molecules because of a shift to the M1 phenotype [63–66]. Thus, Aβ and SNCA seem to be favoring not only the activation of the microglia to the pro-inflammatory phenotype (M1), but also the increase in the responsiveness of the microglia to the harmful stimuli. In both cases, the result will be an over activation of the microglia with the subsequent release of increased levels of pro-inflammatory mediators and ROS, enhancing the damage induced by the initial exposure to Aβ and SNCA [67–69]. Similarly, the chronic exposure to these inflammatory mediators can induce the repolarization of the microglia changing from the M2 to the M1 phenotype [70]. However, additional research is necessary to properly address the significance of microglial priming and the effects of the exposure to different levels of pro-inflammatory stimuli [71]. Indeed, the work conducted by Pourbadie and cols. [72] seems to suggest that low doses of TLR ligands can exert beneficial effects on the neuronal circuitry.

4.5 Mitochondrial dysfunction

Another feature of both pathologies is the affectation of the mitochondrial functionality. Both Aβ and SNCA have the ability to interact with this critical organelle. While Aβ has been detected outside and inside the mitochondria being able to directly induce the several mitochondria-related apoptotic pathways, such as the B-cell lymphoma 2 (BCL2)-beclin1 (BECN1) complex [73, 74], SNCA can induce the activation of the mitochondrial membrane permeability transition pore, promoting mitochondrial swelling and leading to mitochondrial degradation. Indeed, when SNCA degradation is blocked by means of proteasome inhibition, mitochondria result as one of the first organelles to be affected. Moreover, TOM40, a protein that is part of the mitochondrial import machinery, has proven to be determinant of the SNCA-mediated mitochondrial failure [75–77]. Importantly, one of the most critical end points of the mitochondrial failure is the increased production of ROS which is able to induce the activation of microglia and astrocytes as well as to trigger the inflammatory response mediated by the TLRs. Vice versa, the increased levels of pro-inflammatory mediators, such as TNF-α, induced by the activation of the TLRs by means of the Aβ and SNCA can also lead to mitochondrial dysfunction mainly through mitochondrial fragmentation [78].

5. Aging and neuroinflammation: self-conditioning to autodestruction?

Although aging constitutes a natural process, it has been considered from long ago as the main factor for several age-related conditions. However, we must realize that even when aging implies the progressive decay of several biological systems, the main issue with aging is the time span of exposure to different exogenous and potentially harmful stimuli (http://www.iarc.fr) [79, 80]. If we include the genetic and epigenetic heterogeneity between subjects as another factor to consider, it is almost evident that the aging process will follow different pathways depending on the particularities of each subject [81–83].

As previously mentioned, several works have evidenced the link between aging and neurodegenerative disorders. In the context of neuroinflammation, the immune system decay and a pro-inflammatory status are part of the aging process. Because of the increased levels of circulating inflammatory cytokines and the impaired performance of the cellular components involved in the immune response, a chronic exposure to an inflammatory environment verifies for all the biological systems. Regrettably, it has been demonstrated that the brain exhibits the same age-related pro-inflammatory deviation [84–87]. This general inflammatory status of the brain is currently termed as inflammaging.

In general terms, inflammaging is defined by the loss of the inflammatory homeostasis shifting to a pro-inflammatory condition with aging as the determinant factor. Moreover, it has been evidenced that inflammaging is caused by the deregulated function of the inflammasomes, the intracellular structures where several pro-inflammatory mediators are synthetized including several cytokines [84, 88–90]. Moreover, some works have also suggested that inflammaging involves not the deregulation of TLR expression, but the signal cascades triggered after its activation through different microRNAs [91].

On the other hand, inflammaging can also relate with cell senescence. Regrettably, cell senescence also verifies in the immune system and affects the immune cells of both the peripheral system and the CNS. Although astrocytes are believed to be the only cells able to express senescence markers, different researches have evidenced that microglia also exhibit several age-related morphological and biochemical changes. Indeed, the increased levels of activation markers including the cluster of differentiation 11b, 11c, and 14, along with the increased production of TNF-α, IL-1β, IL-6, and reactive oxygen species (ROS) allow to dimension the effect of senescence on the physiology of the immunocompetent cells within the brain [92].

6. Concluding remarks

Inflammatory milieu is an extremely complex event. Moreover, it becomes even more complicated when we introduce the neurodegenerative process as part of the inflammatory equation. In this case, the final outcome will not only be determined by the production and release of the pro-inflammatory mediators and the specific responses triggered in the different cell types present in the brain, but it will also depend on the physiological status of these cells. Aging, and the differential exposure to xenobiotics, will certainly determine the health status of the cells and its ability to answer properly to the inflammatory stimulus and to resist a pro-inflammatory condition. At the basis of all these processes, the molecular mechanisms triggered by the TLRs play a critical role during both AD and PD establishment and progression. Moreover, through the cross talk with additional signaling pathways, TLR cascade is able to interfere with different aspects of the cell physiology from energy production to cytoskeleton rearrangements. On the other hand, less is known regarding other representatives of the TLR family and their impact on AD/PD pathophysiology. For example, some evidence seems to suggest that while TLR9 will exert a protective effect in the context of the neurodegenerative process driven by Aβ and SNCA, TLR3 will also enhance the release of pro-inflammatory mediators [93].

To date, significant evidence seems to confirm the key role of the inflammatory milieu in the neurodegenerative process and this situation should prompt researchers to increase their efforts to understand this cascade of events and to unveil the missing points of an inflammatory-based hypothesis of the neurodegenerative disorders.

Conflicts of interest

The authors declare no competing interest regarding the publication of this work.

Funding

This work was supported by grants from the Basal Center of Excellence in Aging and Regeneration AFB 170005 and FONDECYT (no. 1160724) to NCI and by FONDECYT N° 11170212 to SBC.

Author details

Juan M. Zolezzi[1], Sussy Bastías-Candia[1] and Nibaldo C. Inestrosa[1,2,3]*

1 Faculty of Biological Sciences, Department of Cellular and Molecular Biology, Center for Aging and Regeneration (CARE Chile UC), P. Catholic University of Chile, Santiago, Chile

2 Faculty of Medicine, School of Psychiatry, Centre for Healthy Brain Ageing, University of New South Wales, Sydney, Australia

3 Center for Excellence in Biomedicine of Magellan (CEBIMA), University of Magallanes, Punta Arenas, Chile

*Address all correspondence to: ninestrosa@bio.puc.cl

References

[1] Alzheimer's Disease International. World Alzheimer Report. 2016. Available from: https://www.alz.co.uk/research/world-report [Accessed: December 20, 2018]

[2] Heneka MT, Golenbock DT, Latz E. Innate immunity in Alzheimer's disease. Nature Immunology. 2015;**16**:229-236. DOI: 10.1038/ni.3102

[3] Heneka MT, Carson MJ, El Khoury J, Landreth GE, Brosseron F, Feinstein DL, et al. Neuroinflammation in Alzheimer's disease. Lancet Neurology. 2015;**14**:388-405. DOI: 10.1016/S1474-4422(15)70016-5

[4] Ardura-Fabregat A, Boddeke EWGM, Boza-Serrano A, Brioschi S, Castro-Gomez S, Ceyzériat K, et al. Targeting neuroinflammation to treat Alzheimer's disease. CNS Drugs. 2017;**12**:1057-1082. DOI: 10.1007/s40263-017-0483-3

[5] Abbott NJ, Patabendige AA, Dolman DE, Yusof SR, Begley DJ. Structure and function of the blood-brain barrier. Neurobiology of Disease. 2010;**37**:13-25. DOI: 10.1016/j.nbd.2009.07.030

[6] Zolezzi JM, Inestrosa NC. Peroxisome proliferator-activated receptors and Alzheimer's disease: Hitting the blood–brain barrier. Molecular Neurobiology. 2013;**48**:438-451. DOI: 10.1007/s12035-013-8435-5

[7] Nation DA, Sweeney MD, Montagne A, Sagare AP, D'Orazio LM, Pachicano M, et al. Blood-brain barrier breakdown is an early biomarker of human cognitive dysfunction. Nature Medicine. 2019;**25**:270-276. DOI: 10.1038/s41591-018-0297-y

[8] Ransohoff R, Brown M. Innate immunity in the central nervous system. Journal of Clinical Investigation. 2012;**122**:1164-1171. DOI: 10.1172/JCI58644

[9] Strle K, Zhou JH, Shen WH, Broussard SR, Johnson RW, Freund GG, et al. Interleukin-10 in the brain. Critical Reviews in Immunology. 2001;**21**:427-449. DOI: 10.1615/CritRevImmunol.v21.i5.20

[10] Malipiero U, Koedel U, Pfister HW, Levéen P, Bürki K, Reith W, et al. TGFbeta receptor II gene deletion in leucocytes prevents cerebral vasculitis in bacterial meningitis. Brain. 2006;**129**:2404-2415. DOI: 10.1093/brain/awl192

[11] Landreth G, Reed-Geaghan E. Toll-like receptors in Alzheimer's disease. Current Topics in Microbiology and Immunology. 2009;**336**:137-153. DOI: 10.1007/978-3-642-00549-7_8

[12] Hanke M, Kielian T. Toll-like receptors in health and disease in the brain: Mechanisms and therapeutic potential. Clinical Science (London). 2011;**121**:367-387. DOI: 10.1042/CS20110164

[13] Takeuchi O, Akira S. Pattern recognition receptors and inflammation. Cell. 2010;**140**:805-820. DOI: 10.1016/j.cell.2010.01.022

[14] Piccinini AM, Midwood KS. DAMPening inflammation by modulating TLR signaling. Mediators of Inflammation. 2010;**2010**:pii:672395. DOI: 10.1155/2010/672395

[15] Bianchi ME. DAMPs, PAMPs and alarmins: All we need to know about danger. Journal of Leukocyte Biology. 2007;**81**:1-5. DOI: 10.1189/jlb.0306164

[16] Atmaca HT, Kul O, Karakus E, Terzi OS, Canpolat S, Anteplioglu T. Astrocytes, microglia/macrophages, and neurons expressing toll-like receptor

11 contribute to innate immunity against encephalitic toxoplasma gondii infection. Neuroscience. 2014;**269**:184-191. DOI: 10.1016/j.neuroscience.2014.03.049

[17] Mishra BB, Mishra PK, Teale JM. Expression and distribution of toll-like receptors in the brain during murine neurocysticercosis. Journal of Neuroimmunology. 2006;**181**:46-56. DOI: 10.1016/j.jneuroim.2006.07.019

[18] Kawai T, Akira S. Toll-like receptor downstream signaling. Arthritis Research and Therapy. 2005;7:12-19. DOI: 10.1186/ar1469

[19] Kawai T, Akira S. Signaling to NF-κB by toll-like receptors. Trends in Molecular Medicine. 2007;**13**:460-469. DOI: 10.1016/j.molmed.2007.09.002

[20] Ishitani T, Ninomiya-Tsuji J, Nagai S, Nishita M, Meneghini M, Barker N, et al. The TAK1-NLK-MAPK-related pathway antagonizes signaling between β-catenin and transcription factor TCF. Nature. 1999;**399**:798-802. DOI: 10.1038/21674

[21] Zhang ZY, Li SZ, Zhang HH, Wu QR, Gong J, Liang T, et al. Stabilization of ATF5 by TAK1-Nemo-like kinase critically regulates the interleukin-1β-stimulated C/EBP signaling pathway. Molecular and Cellular Biology. 2015;**35**:778-788. DOI: 10.1128/MCB.01228-14

[22] Monick MM, Carter AB, Robeff PK, Flaherty DM, Peterson MW, Hunninghake GW. Lipopolysaccharide activates Akt in human alveolar macrophages resulting in nuclear accumulation and transcriptional activity of beta-catenin. Journal of Immunology. 2001;**166**:4713-4720. DOI: 10.4049/jimmunol.166.7.4713

[23] Arbibe L, Mira JP, Teusch N, Kline L, Guha M, Mackman N, et al. Toll-like receptor 2-mediated NF-kappa B activation requires a Rac1-dependent pathway. Nature Immunology. 2000;**1**:533-540. DOI: 10.1038/82797

[24] Luu K, Greenhill CJ, Majorors A, Decker T, Jenkins BJ, Mansell A. STAT1 plays a role in TLR signal transduction and inflammatory responses. Immunology and Cell Biology. 2014;**92**:761-769. DOI: 10.1038/icb.2014.51

[25] Song F, Zeng K, Liao L, Yu Q, Tu P, Wang X. Schizadrin a inhibits microglia-mediated neuroinflammation through inhibiting TRAF6-NF-κB and Jak-Stat3 signaling pathways. PLoS One. 2016;**11**:1-16. DOI: 10.1371/journal.pone.0149991

[26] Selkoe DJ. Alzheimer's disease. Cold Spring Harbor Perspectives in Biology. 2011;**3**:a004457. DOI: 10.1101/cshperspect.a004457

[27] Selkoe DJ, Hardy J. The amyloid hypothesis of Alzheimer's disease at 25 years. EMBO Molecular Medicine. 2016;**8**:595-608. DOI: 10.15252/emmm.201606210

[28] Singh I, Sagare AP, Coma M, Perlmutter D, Gelein R, Bell RD, et al. Low levels of copper disrupt brain amyloid-β homeostasis by altering its production and clearance. Proceedings of the National Academy of Sciences of the United States of America. 2013;**110**:14771-14776. DOI: 10.1073/pnas.1302212110

[29] Yan R, Vassar R. Targeting the β secretase BACE1 for Alzheimer's disease therapy. Lancet Neurology. 2014;**13**:319-329. DOI: 10.1016/S1474-4422(13)70276-X

[30] Zolezzi JM, Bastías-Candia S, Santos MJ, Inestrosa NC. Alzheimer's disease: Relevant molecular and physiopathological events affecting amyloid-β brain balance and the putative role of PPARs. Frontiers in Aging Neuroscience. 2014;**6**:176. DOI: 10.3389/fnagi.2014.00176

[31] Salminen A, Ojala J, Kauppinen A, Kaarniranta K, Suuronen T. Inflammation in Alzheimer's disease: Amyloid-beta oligomers trigger innate immunity defense via pattern recognition receptors. Progress in Neurobiology. 2009;**87**:181-194. DOI: 10.1016/j.pneurobio.2009.01.001

[32] Liu S, Liu Y, Hao W, Wolf L, Kiliaan AJ, Penke B, et al. TLR2 is a primary receptor for Alzheimer's amyloid b peptide to trigger neuroinflammatory activation. Journal of Immunology. 2012;**188**:1098-1107. DOI: 10.4049/jimmunol.1101121

[33] Fassbender K, Walter S, Kühl S, Landmann R, Ishii K, Bertsch T, et al. The LPS receptor (CD14) links innate immunity with Alzheimer's disease. FASEB Journal. 2004;**18**:203-205. DOI: 10.1096/fj.03-0364fje

[34] Iwashyna TJ, Ely EW, Smith DM, Langa KM. Long-term cognitive impairment and functional disability among survivors of severe sepsis. Journal of the American Medical Association. 2010;**304**:1787-1794. DOI: 10.1001/jama.2010.1553

[35] McManus RM, Heneka MT. Role of neuroinflammation in neurodegeneration: New insights. Alzheimer's Research & Therapy. 2017;**9**:14. DOI: 10.1186/s13195-017-0241-2

[36] Dunn N, Mullee M, Perry VH, Holmes C. Association between dementia and infectious disease: Evidence from a case–control study. Alzheimer Disease and Associated Disorders. 2005;**19**:91-94. DOI: 10.1097/01.wad.0000165511.52746.1f

[37] Holmes C, El-Okl M, Williams AL, Cunningham C, Wilcockson D, Perry VH. Systemic infection, interleukin 1beta, and cognitive decline in Alzheimer's disease. Journal of Neurology, Neurosurgery and Psychiatry. 2003;**74**:788-789. DOI: 10.1136/jnnp.74.6.788

[38] Holmes C, Cunningham C, Zotova E, Woolford J, Dean C, Kerr S, et al. Systemic inflammation and disease progression in Alzheimer disease. Neurology. 2009;**73**:768-774. DOI: 10.1212/WNL.0b013e3181b6bb95

[39] Kalia LV, Lang AE. Parkinson's disease. Lancet. 2015;**386**:896-912. DOI: 10.1016/S0140-6736(14)61393-3

[40] Murphy DD, Rueter SM, Trojanowski JQ, Lee VM. Synucleins are developmentally expressed, and alpha-synuclein regulates the size of the presynaptic vesicular pool in primary hippocampal neurons. The Journal of Neuroscience. 2000;**20**:3214-3220. DOI: 10.1523/JNEUROSCI.20-09-03214.2000

[41] Cabin DE, Shimazu K, Murphy D, Cole NB, Gottschalk W, McIlwain KL, et al. Synaptic vesicle depletion correlates with attenuated synaptic responses to prolonged repetitive stimulation in mice lacking alpha-synuclein. The Journal of Neuroscience. 2002;**22**:8797-8807. DOI: 10.1523/JNEUROSCI.22-20-08797.2002

[42] Larsen KE, Schmitz Y, Troyer MD, Mosharov E, Dietrich P, Quazi AZ, et al. Alpha-synuclein overexpression in PC12 and chromaffin cells impairs catecholamine release by interfering with a late step in exocytosis. The Journal of Neuroscience. 2006;**26**:11915-11922. DOI: 10.1523/JNEUROSCI.3821-06.2006

[43] Li YH, Gao N, Ye YW, Li X, Yu S, Yang H, et al. Alpha-synuclein functions as a negative regulator for expression of tyrosine hydroxylase. Acta Neurologica Belgica. 2011;**111**:130-135

[44] Butler B, Goodwin S, Saha K, Becker J, Sambo D, Davari P, et al. Dopamine transporter activity is modulated by alpha-synuclein.

The Journal of Biological Chemistry. 2015;**290**:29542-29554. DOI: 10.1074/jbc.M115.691592

[45] Jia SH, Li K, Su W, Li SH, Chen HB. Impairment in the intention formation and execution phases of prospective memory in Parkinson's disease. Frontiers in Neuroscience. 2018;**12**:98. DOI: 10.3389/fnins.2018.00098

[46] Hawkes CH, Del Tredici K, Braak H. A timeline for Parkinson's disease. Parkinsonism & Related Disorders. 2010;**16**:79-84. DOI: 10.1016/j.parkreldis.2009.08.007

[47] Bridi JC, Hirth F. Mechanisms of α-synuclein induced synaptopathy in Parkinson's disease. Frontiers in Neuroscience. 2018;**12**:80. DOI: 10.3389/fnins.2018.00080

[48] Bastías-Candia S, Di Benedetto M, D'Addario C, Candeletti S, Romualdi P. Combined exposure to agriculture pesticides, paraquat and maneb, induces alterations in the N/OFQ-NOPr and PDYN/KOPr systems in rats: Relevance to sporadic Parkinson's disease. Environmental Toxicology. 2015;**30**:656-663. DOI: 10.1002/tox.21943

[49] Stefanova N, Fellner L, Reindl M, Masliah E, Poewe W, Wenning GK. Toll-like receptor 4 promotes α-synuclein clearance and survival of nigral dopaminergic neurons. The American Journal of Pathology. 2011;**179**:954-963. DOI: 10.1016/j.ajpath.2011.04.013

[50] Fellner L, Irschick R, Schanda K, Reindl M, Klimaschewski L, Poewe W, et al. Toll-like receptor 4 is required for α-synuclein dependent activation of microglia and astroglia. Glia. 2013;**61**:349-360. DOI: 10.1002/glia.22437

[51] Kim C, Ho DH, Suk JE, You S, Michael S, Kang J, et al. Neuron-released oligomeric α-synuclein is an endogenous agonist of TLR2 for paracrine activation of microglia. Nature Communications. 2013;**4**:1562. DOI: 10.1038/ncomms2534

[52] Sanchez-Guajardo V, Tentillier N, Romero-Ramos M. The relation between α-synuclein and microglia in Parkinson's disease: Recent developments. Neuroscience. 2015;**302**:47-58. DOI: 10.1016/j.neuroscience.2015.02.008

[53] Ganguly U, Chakrabarti SS, Kaur U, Mukherjee A, Chakrabarti S. Alpha-synuclein, proteotoxicity and Parkinson's disease: Search for neuroprotective therapy. Current Neuropharmacology. 2017;**15**. DOI: 10.2174/1570159X15666171129100944

[54] Lee HJ, Suk JE, Patrick C, Bae EJ, Cho JH, Rho S, et al. Direct transfer of alpha-synuclein from neuron to astroglia causes inflammatory responses in synucleinopathies. The Journal of Biological Chemistry. 2010;**285**:9262-9272. DOI: 10.1074/jbc.M109.081125

[55] Bassil F, Fernagut PO, Bezard E, Pruvost A, Leste-Lasserre T, Hoang QQ, et al. Reducing C-terminal truncation mitigates synucleinopathy and neurodegeneration in a transgenic model of multiple system atrophy. Proceedings of the National Academy of Sciences of the United States of America. 2016;**113**:9593-9598. DOI: 10.1073/pnas.1609291113

[56] Wang W, Nguyen LT, Burlak C, Chegini F, Guo F, Chataway T, et al. Caspase-1 causes truncation and aggregation of the Parkinson's disease-associated protein α-synuclein. Proceedings of the National Academy of Sciences of the United States of America. 2016;**113**:9587-9592. DOI: 10.1073/pnas.1610099113

[57] Reed-Geaghan EG, Savage JC, Hise AG, Landreth GE. CD14 and

toll-like receptors 2 and 4 are required for fibrillar Ab-stimulated microglial activation. Journal of Neuroscience. 2009;**29**:11982-11992. DOI: 10.1523/JNEUROSCI.3158-09.2009

[58] Sweeney MD, Sagare AP, Zlokovic BV. Blood-brain barrier breakdown in Alzheimer disease and other neurodegenerative disorders. Nature Reviews Neurology. 2018;**14**:133-150. DOI: 10.1038/nrneurol.2017.188

[59] Yirmiya R, Goshen I. Immune modulation of learning, memory, neural plasticity and neurogenesis. Brain, Behavior, and Immunity. 2011;**25**: 181-213. DOI: 10.1016/j.bbi.2010.10.015

[60] Turrigiano GG. The self-tuning neuron: Synaptic scaling of excitatory synapses. Cell. 2008;**135**:422-435. DOI: 10.1016/j.cell.2008.10.008

[61] Stellwagen D, Malenka RC. Synaptic scaling mediated by glial TNF-alpha. Nature. 2006;**440**:1054-1059. DOI: 10.1038/nature04671

[62] Riazi K, Galic MA, Kuzmiski JB, Ho W, Sharkey KA, Pittman QJ. Microglial activation and TNFalpha production mediate altered CNS excitability following peripheral inflammation. Proceedings of the National Academy of Sciences of the United States of America. 2008;**105**: 17151-17156. DOI: 10.1073/pnas.0806682105

[63] Ajami B, Bennett JL, Krieger C, et al. Local self-renewal can sustain CNS microglia maintenance and function throughout adult life. Nature Neuroscience. 2007;**10**:1538-1543. DOI: 10.1038/nn2014

[64] Lynch MA. The multifaceted profile of activated microglia. Molecular Neurobiology. 2009;**40**:139-156. DOI: 10.1007/s12035-009-8077-9

[65] Perry VH. Contribution of systemic inflammation to chronic neurodegeneration. Acta Neuropathologica. 2010;**120**:277-286. DOI: 10.1007/s00401-010-0722-x

[66] Ghosh S, Wu MD, Shaftel SS, et al. Sustained interleukin-1beta overexpression exacerbates tau pathology despite reduced amyloid burden in an Alzheimer's mouse model. Journal of Neuroscience. 2013;**33**:5053-5064. DOI: 10.1523/JNEUROSCI.4361-12.2013

[67] Rojanathammanee L, Floden AM, Manocha GD, et al. Attenuation of microglial activation in a mouse model of Alzheimer's disease via NFAT inhibition. Journal of Neuroinflammation. 2015;**12**:42. DOI: 10.1186/s12974-015-0255-2

[68] Wu Z, Sun L, Hashioka S, et al. Differential pathways for interleukin-1beta production activated by chromogranin a and amyloid beta in microglia. Neurobiology of Aging. 2013;**34**:2715-2725. DOI: 10.1016/j.neurobiolaging.2013.05.018

[69] Rocha EM, Smith GA, Park E, et al. Sustained systemic glucocerebrosidase inhibition induces brain alpha-synuclein aggregation, microglia and complement C1q activation in mice. Antioxidants and Redox Signaling. 2015;**23**:550-564. DOI: 10.1089/ars.2015.6307

[70] Perry VH, Teeling J. Microglia and macrophages of the central nervous system: The contribution of microglia priming and systemic inflammation to chronic neurodegeneration. Seminars in Immunopathology. 2013;**35**:601-612. DOI: 10.1007/s00281-013-0382-8

[71] Caldeira C, Cunha C, Vaz AR, Falcão AS, Barateiro A, Seixas E, et al. Key aging-associated alterations in primary microglia response to beta-amyloid stimulation. Frontiers in Aging Neuroscience. 2017;**9**:277. DOI: 10.3389/fnagi.2017.00277

[72] Pourbadie HG, Sayyah M, Khoshkholgh-Sima B, Choopani S, Nategh M, Motamedi F, et al. Early minor stimulation of microglial TLR2 and TLR4 receptors attenuates Alzheimer's disease-related cognitive deficit in rats: Behavioral, molecular, and electrophysiological evidence. Neurobiology of Aging. 2018;**70**:203-216. DOI: 10.1016/j.neurobiolaging.2018.06.020

[73] Cai Q, Tammineni P. Mitochondrial aspects of synaptic dysfunction in Alzheimer's disease. Journal of Alzheimer's Disease. 2017;**57**:1087-1103. DOI: 10.3233/JAD-160726

[74] Kerr JS, Adriaanse BA, Greig NH, Mattson MP, Cader MZ, Bohr VA, et al. Mitophagy and Alzheimer's disease: Cellular and molecular mechanisms. Trends in Neurosciences. 2017;**40**:151-166. DOI: 10.1016/j.tins.2017.01.002

[75] Banerjee K, Sinha M, Pham CLL, Jana S, Chanda D, Cappai R, et al. Alpha-synuclein induced membrane depolarization and loss of phosphorylation capacity of isolated rat brain mitochondria: Implications in Parkinson's disease. FEBS Letters. 2010;**584**:1571-1576. DOI: 10.1111/jnc.12966

[76] Bender A, Desplats P, Spencer B, et al. TOM40 mediates mitochondrial dysfunction induced by α-synuclein accumulation in Parkinson's disease. PLoS One. 2013;**8**:e62277. DOI: 10.1371/journal.pone.0062277

[77] Junn E, Mouradian MM. Human α-Synuclein over-expression increases intracellular reactive oxygen species levels and susceptibility to dopamine. Neuroscience Letters. 2002;**320**:146-150. DOI: 10.1016/S0304-3940(02)00016-2

[78] Wu B, Li J, Ni H, Zhuang X, Qi Z, Chen Q, et al. TLR4 activation promotes the progression of experimental autoimmune myocarditis to dilated cardiomyopathy by inducing mitochondrial dynamic imbalance. Oxidative Medicine and Cellular Longevity. 2018;**2018**:3181278. DOI: 10.1155/2018/3181278

[79] Liu Y-Z, Wang Y-X, Jiang C-L. Inflammation: The common pathway of stress-related diseases. Frontiers in Human Neuroscience. 2017;**11**:316. DOI: 10.3389/fnhum.2017.00316

[80] McCormick CM, Hodges TE. Stress, glucocorticoids, and brain development in rodent models. In: Fink G, editor. Stress: Neuroendocrinology and Neurobiology. San Diego: Academic Press; 2017. pp. 197-206

[81] Ashapkin VV, Kutueva LI, Vanyushin BF. Aging as an epigenetic phenomenon. Current Genomics. 2017;**18**:385-407. DOI: 10.2174/13892029 18666170412112130

[82] Masser DR, Hadad N, Porter HL, Mangold CA, Unnikrishnan A, Ford MM, et al. Sexually divergent DNA methylation patterns with hippocampal aging. Aging Cell. 2017;**16**:1342-1352. DOI: 10.1111/acel.12681

[83] Lunnon K, Smith R, Hannon E, De Jager PL, Srivastava G, Volta M, et al. Methylomic profiling implicates cortical deregulation of ANK1 in Alzheimer's disease. Nature Neuroscience. 2014;**17**:1164-1170. DOI: 10.1038/nn.3782

[84] Cao W, Zheng H. Peripheral immune system in aging and Alzheimer's disease. Molecular Neurodegeneration. 2018;**1**:51. DOI: 10.1186/s13024-018-0284-2

[85] Schwalm MT, Pasquali M, Miguel SP, Dos Santos JP, Vuolo F, Comim CM, et al. Acute brain inflammation and oxidative damage are related to long-term cognitive deficits and markers of neurodegeneration in sepsis-survivor rats. Molecular Neurobiology.

2014;**49**:380-385. DOI: 10.1007/s12035-013-8526-3

[86] Wang LM, Wu Q, Kirk RA, Horn KP, Ebada Salem AH, Hoffman JM, et al. Lipopolysaccharide endotoxemia induces amyloid-β and p-tau formation in the rat brain. American Journal of Nuclear Medicine and Molecular Imaging. 2018;**8**:86-99. ISSN:2160-8407/ajnmmi0074312

[87] Wyss-Coray T. Ageing, neurodegeneration and brain rejuvenation. Nature. 2016;**539**:180-186. DOI: 10.1038/nature20411

[88] Giunta B, Fernandez F, Nikolic WV, Obregon D, Rrapo E, Town T, et al. Inflammaging as a prodrome to Alzheimer's disease. Journal of Neuroinflammation. 2008;**5**:51. DOI: 10.1186/1742-2094-5-51

[89] Frasca D, Blomberg BB. Inflammaging decreases adaptive and innate immune responses in mice and humans. Biogerontology. 2016;**17**:7-19. DOI: 10.1007/s10522-015-9578-8

[90] Yao H, Rahman I. Perspectives on translational and therapeutic aspects of SIRT1 in inflammaging and senescence. Biochemical Pharmacology. 2012;**84**:1332-1339. DOI: 10.1016/j.bcp.2012.06.031

[91] Olivieri F, Rippo MR, Prattichizzo F, Babini L, Graciotti L, Recchioni R, et al. Toll like receptor signaling in "inflammaging": microRNA as new players. Immunity & Ageing. 2013;**10**:11. DOI: 10.1186/1742-4933-10-11

[92] Cameron B, Landreth GE. Inflammation, microglia, and Alzheimer's disease. Neurobiology of Disease. 2010;**37**:503-509. DOI: 10.1016/j.nbd.2009.10.006

[93] Walker DG, Tang TM, Lue LF. Increased expression of toll-like receptor 3, an anti-viral signaling molecule, and related genes in Alzheimer's disease brains. Experimental Neurology. 2018;**309**:91-106. DOI: 10.1016/j.expneurol.2018.07.016